Living Seashells
of the Tropical Indo-Pacific

Photographic guide with 1500+ species covered

Andrey Ryanskiy

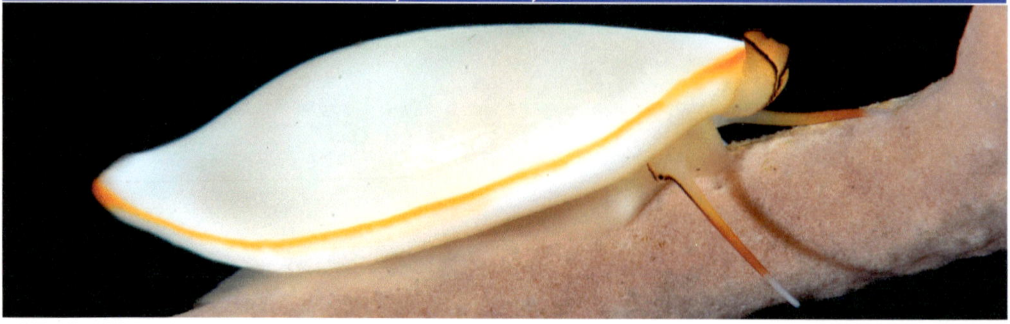

INTRODUCTION Seashell or sea shells are the hard exoskeleton of mollusks such as snails, clams, chitons. For most people, acquaintance with mollusks began with empty shells. These shells often delight the eye with a variety of shapes and colors. Conchology studies the mollusk shells and this science dates back to the 17th century. However, modern science - malacology is the study of mollusks as whole organisms. Today more and more people are interacting with ocean - divers, snorkelers, beach goers - all of them often find in the seas not empty shells, but live mollusks - living shells, whose appearance is significantly different from museum specimens. This book serves as a tool for identifying such animals.

The book covers the region from the Red Sea to Hawaii, Marshall Islands and Guam. Inside the book:
• Photographs of 1500+ species, including one hundred cowries (Cypraeidae) and more than one hundred twenty allied cowries (Ovulidae) of the region;
• Live photo of hundreds of species have never before appeared in field guides or popular books;
• Convenient pictorial guide at the beginning and index at the end of the book.

ACKNOWLEDGMENTS The significant part of photographs in this book were made by Jeanette Johnson and Scott Johnson during the decades of diving and exploring the beautiful reefs of Indo-Pacific from Indonesia and Philippines to Hawaii and Solomons. They provided to readers not only the great photos but also in-depth knowledge of the fascinating world of living seashells.

Sincere thanks to Philippe Bouchet, National Museum of Natural History (Paris), for inviting the author to participate in the La Planete Revisitee expedition program and permission to use some of the NMNH photos.

Special thanks to Alexander Fedosov, Yuri Kantor, Max Marrow, Richard Salisbury, Yasunori Kano, Tsuyoshi Takano, Boris Sirenko, Malcolm Tait Sanders, whose scientific support helped to identify many species from this photo guide. Sue Churchill and Rogan Draper, authors of the brilliant book about ovulids "The Invisibility Cloak", supported a project with deep first-hand knowledge and excellent photos.

I was lucky to receive ongoing support in my book projects from Philippe and Guido Poppe. Jacky and Evelyn Guillot de Suduiraut are kindly acknowledged for permission to use photos of Emmanuel Guillot de Suduiraut. Sven Kahlbrock knows Red Sea ovulids as no one else and his photos enriched the book. Massimo Scali shared with readers his photos of the rare Indian Ocean cowries. Jake Adams, Josina Tivaouane and Jocelyn Senia helped to make Tridacna pages successfully finalized. Bob Abela helped with some rare species, photographed in Guam, Charles E. Rawlings contributed unique volutidae photos from the Indian Ocean. Brian Mayes kindly shared with readers his outstanding collection of superb living ovulidae images.

Special mention must be paid to amazing Sci-Hub project (The first website in the world to provide mass & public access to research papers!), Biodiversity Heritage Library and ResearchGate networking site.

Anke Wentzler, Arthur Anker, Arto Jarvenpaa, Asyraf Hadee, Barbier Jean-Pierre, Bernd Hoppe, Bjarte Hekonsen, Chay Hoon, Christina Sylvester, Dave Johnson, David Burdick, Dawn Goebbels, Dennis Teck Lee Liew, Igor Bogachenko, Irina Sevriuk, Jay Lord, Johji Nishio, Luc Faucompre, Marty Kiser, Nick Hobgood, Philippe Bourjon, Rafi Amar, Ravi Vadher, Ria Tan, Rokus Groeneveld, Richard Parker, Sebastien Vasquez, Petr Vorotnicov, Shawn Miller, Stan Jazwinski, Stephen D.A. Smith, Steven Ng, Tsu Soo Tan, Ular Tikk, Uwe Weinreich, Wolfgang Kaufmann, Yoichi Sato, Yui Kawahara, Yves Thevenet generously shared with the author and readers their living seashells photos.

I would like to thank my wife Irina Khlopunova for patient support during the work on this book.

ABBREVIATIONS

Principal photo contributors: ©JJ: Jeanette Johnson ©SJ: Scott Johnson

IT - Identification Tentative; sp. - used when the actual specific name cannot or need not be specified

cf. - used to indicate undescribed species assumed related to, but distinct from a described species

WIO, IO, IP, IWP, WP, WCP, PO, SP - West Indian Ocean, Indian Ocean, Indo-Pacific, Indo-West Pacific, West Pacific, Western Central Pacific, Pacific Ocean, South Pacific accordingly.

Front cover photos: upper left: *Dentiovula eizoi* (Igor Bogachenko), upper right: *Vexillum jeanetteae* (Jeanette Johnson, yes, it is not a coincidence of names!), bottom left: *Cribrarula cribraria* (Jeanette Johnson), b. right: *Limaria fragilis*.

MARINE GASTROPODS (SNAILS AND SLUGS)

- the most diversified class in the phylum Mollusca
- gastropods mostly have a one-piece, often coiled shell
- they have a well-defined head with two sensory tentacles with eyes
- they have a ventral foot, which gives them their name (stomach and foot, from Greek "gaster" and "pous")
- they are distinguished by an anatomical process known as torsion which occurs in all gastropods during larval development.
- torsion is the rotation of the visceral mass, mantle, and shell 180° with respect to the head and foot
- many gastropods can seal the entrance to their shell with a tough corneous or calcareous operculum
- snails are gastropods with an external shell, slug has no shell or a very reduced shell (in loose terms). We are dealing with marine snails in this book, look for slugs in the "Nudibranchs of the Coral Triangle" book.

BIVALVES - BIVALVIA

- majority of bivalves consist of two identical valves that are connected to one another by a flexible hinge
- bivalves have no head
- some bivalves have eyes
- the gills have evolved into specialized organs for feeding and breathing - ctenidia
- some bivalves have a retractable foot, enabling the animal to burrow in the sand or to move, often with jumps
- bivalves range in size from about one millimeter to over a meter in length
- bivalves are mostly filter feeders, they lack the radular rasping organ typical of most mollusks
- in giant clams (subfamily Tridacninae) symbiotic zooxanthellae are kept within the mantle tissue, providing additional nutrition
- near three million metric tons of bivalves are harvested throughout the world each year
- bivalves were rare in the oceans before Permian–Triassic extinction event (approximately 252 million years ago) but became numerous and diverse after it

CHITONS - POLYPLACOPHORA

- the chitons are exclusively marine, unlike gastropods and bivalves
- all living chitons have eight shell plates surrounded by chitinous girdle and large adhesive foot
- it allows chitons to roll into a protective ball when dislodged
- live in intertidal/subtidal waters along rocky coastlines worldwide
- most chitons are herbivorous grazers, with some omnivorous and some carnivorous species
- chitons are slowly moving animals, but several species can move much faster

PICTORIAL INDEX TO FAMILIES

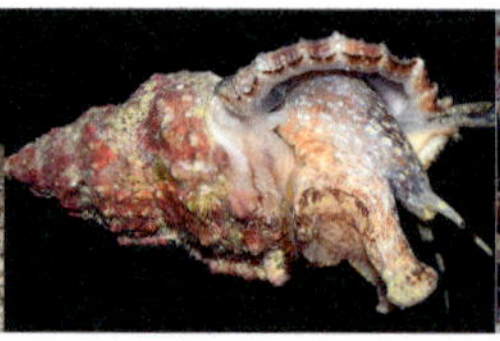

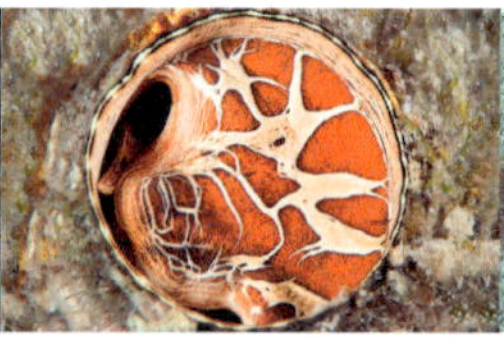

PICTORIAL INDEX TO FAMILIES

PISANIIDAE - 121

FASCIOLARIIDAE - 124

COLUBRARIIDAE - 127

COLUMBELLIDAE - 128

CONIDAE - 134

TEREBRIDAE - 140

CLAVATULIDAE - 156

TURRIDAE - 156

CLATHURELLIDAE - 159

DRILLIIDAE - 161

HORAICLAVIDAE - 164

MANGELIIDAE - 165

MITROMORPHIDAE - 168

RAPHITOMIDAE - 169

PSEUDOMELATOMIDAE - 173

MITRIDAE - 173

MURICIDAE - 184

MURICIDAE - 192

HARPIDAE - 196

OLIVIDAE - 198

ANCILLARIIDAE - 200

COSTELLARIIDAE - 201

TURBINELLIDAE - 214

CYSTISCIDAE - 214

MARGINELLIDAE - 215

CANCELLARIIDAE - 217

VOLUTIDAE - 217

NERITIDAE - 219

PHENACOLEPADIDAE - 219

NERITOPSIDAE - 219

CALLIOSTOMATIDAE - 220

PHASIANELLIDAE - 220

PICTORIAL INDEX TO FAMILIES

TEGULIDAE - 221

TURBINIDAE - 222

TROCHIDAE - 223

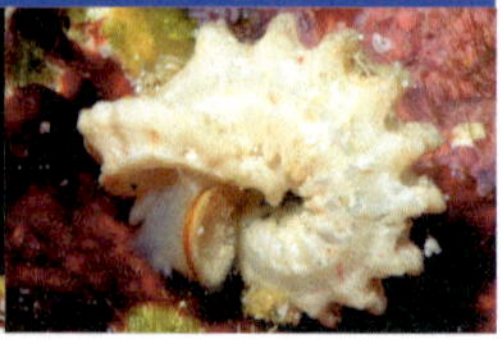
LIOTIIDAE - 228

ANGARIIDAE - 228

HALIOTIDAE - 228

FISSURELLIDAE - 229

CHILODONTAIDAE - 232

ARCHITECTONICIDAE - 233

PYRAMIDELLIDAE - 235

ONCHIDIIDAE - 236

TRIDACNINAE - 237

CARDIIDAE - 242

GALEOMMATIDAE - 244

VENERIDAE - 247

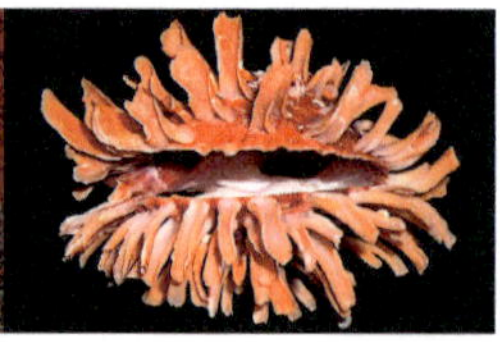
CHAMIDAE - 248

CLAVAGELLIDAE - 249

ARCIDAE - 249

LIMIDAE - 250

LUCINIDAE - 252

MYTILIDAE - 252

GRYPHAEIDAE - 252

OSTREIDAE - 253

PTERIIDAE - 253

PINNIDAE - 255

PECTINIDAE - 256

SPONDYLIDAE - 259

ACANTHOCHITONIDAE - 261

CHITONIDAE - 261

CRYPTOPLACIDAE - 262

ISCHNOCHITONIDAE - 263

SCHIZOCHITONIDAE - 263

BASIC MORPHOLOGY
TECHNICAL TERMS

GASTROPODA

Ovula (Cypraeoidae)

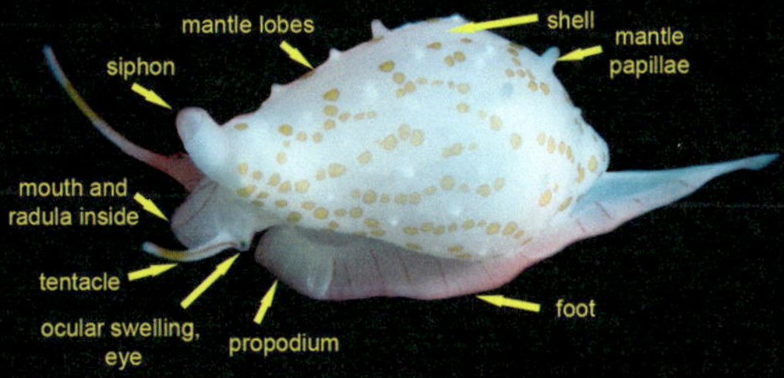

Triton (Tonnoidea)

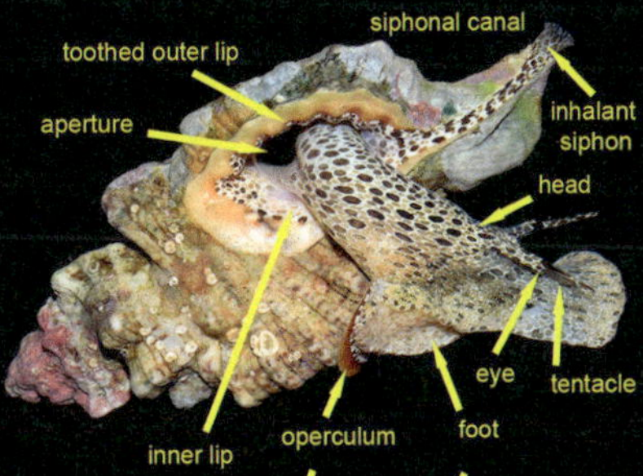

Murex (Muricidae)

Bursa (Bursidae)

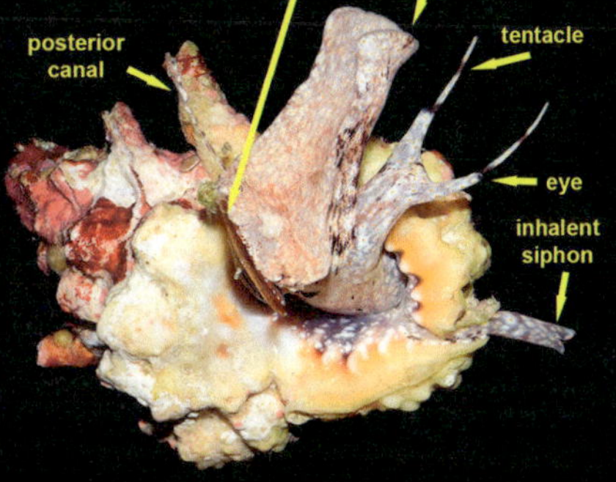

BIVALVIA

Scallops (Pectinidae)

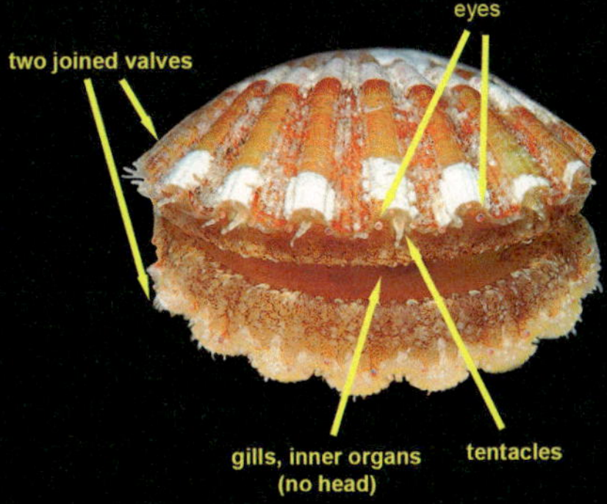

CHITONS - POLYPLACOPHORA

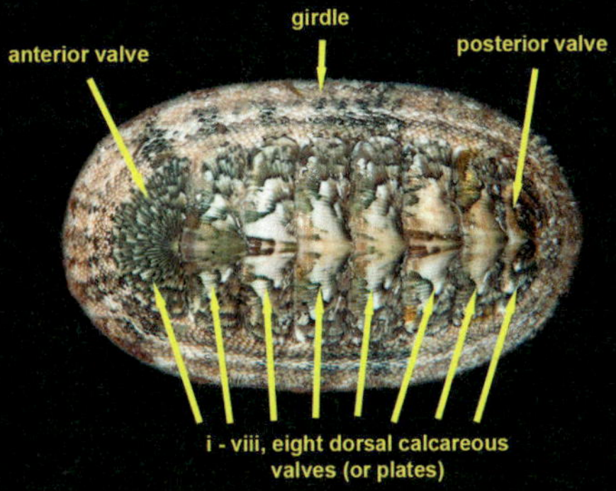

Teulere's Cowry *Barycypraea teulerei* IO: Oman, intertidal mud flats, 67 mm. Transparent mantle with no papillae. Considered extinct due to over-collection, until Massimo Scali found a new population in 2012 at Masirah.

Teulere's Cowry *Barycypraea teulerei* (cont'd) uses shells to nest its eggs. All 3 photos: © Massimo Scali.

Maria's Cowry *Annepona mariae* IP, 21 mm. White with rounded yellow to brown spots. ©JJ.

Hundred-Eyed Cowry *Arestorides argus* IP: East Africa to Marshall Islands, 120 mm. Brownish with wide darker bands and brown ocelli, mantle with raised papillae and depressions that resemble sponge. ©JJ.

Dark-Blotched Cowry *Bistolida fuscomaculata* WCP, 17 mm. Shell with distinct brown blotches. ©JJ.

Swallow Cowry *Bistolida hirundo* IWP, 24 mm. Anterior white transverse mark in the form of a swallow.

Kiener's Cowry *Bistolida kieneri* IWP, 24 mm. Similar in appearance to *B. hirundo* and *B. ursellus*, but anterior white transverse band is irregular in shape and crosses all the grey-bluish area.

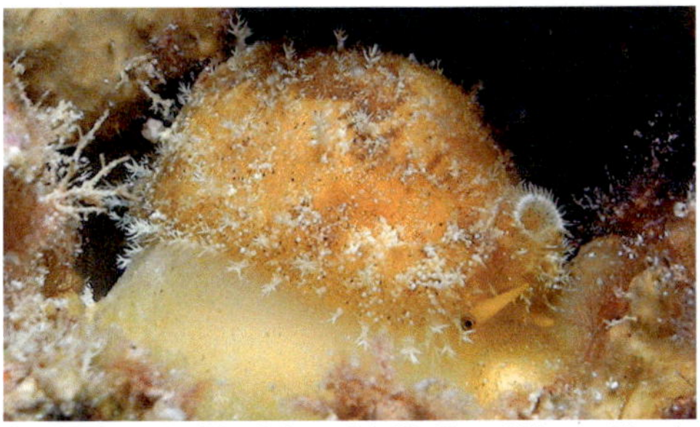

Stolid Cowrie *Bistolida stolida* IP: East Africa to Marshall Islands, 40 mm. Shell with large brown blotches. Yellow-brown mantle with thin white tree-shaped papillae. Feeds on sponges and algae. ©SJ.

Little Bear Cowrie *Bistolida ursellus* IWP to Marshall Isl, 21 mm. Similar in appearance to *B. hirundo* and *B. kieneri*, but the anterior white dorsal band often shaped like a "W" or just like a white blob.

Palish Cowry *Blasicrura pallidula* WP: Philippines to New Caledonia and Samoa, 32 mm. Brownish reticulated mantle with darker and brighter areas and scattered papillae. Left: © Yves Trenevet, right: ©SJ.

Golden Cowry *Callistocypraea aurantium* WCP, 121 mm. Large nocturnal species with yellow or orange shell. Semi-translucent mantle with areas of brown reticulation. Feeds on a grey sponge. ©JJ.

Turtle Cowry *Chelycypraea testudinaria* IP, 153 mm. Oval or sub-cylindrical shell is yellowish, covered with irregular brown spots. mantle is brownish grey with darker longitudinal lines. Photos: ©JJ (left) and ©SJ (right).

Pretty Cowry *Ficadusta pulchella* IWP, 56 mm. © Philippe & Guido Poppe www.poppe-images.com

Walker's Cowry *Contradusta walkeri* IP, 56 mm. Shell with a brown transverse band. © Ria Tan.

Breger's Cowry *Contradusta bregeriana* WCP, 34 mm. Similar in appearance to *C. walkeri* but tends to be smaller, brown transverse band generally less distinct. Right - © Yves Thevenet.

Tan and White Cowry *Cribrarula cribraria* IWP, 35 mm. Shell brown or orange with round white spots, bright orange to red mantle with tapering papillae. Found under shallow rocks, where it feeds on red encrusting sponge. ©JJ.

Gaskoin's Cowry *Cribrarula gaskoini* Hawaii, possibly Marshalls, 30 mm. Orange shell with pale round spots. Orange mantle is densely covered with tapering papillae, feeds on red encrusting sponge, nocturnal. ©SJ.

Pale Sieve Cowry *Cribrarula gaspardi* WCP: Marshalls to Guam, 22 mm. Similar in appearance to *C. cribraria*, but generally smaller and paler, found on the outer slopes in the night. ©JJ.

Dillwyn's Cowry *Cryptocypraea dillwyni* WCP: Indonesia to Micronesia and Polynesia, 17 mm. Pale yellow to yellow-orange with white spots shell, mantle with long papillae. ©JJ.

Panther Cowry *Cypraea pantherina* IO: Red Sea, 120 mm. Shell with brown spots, mantle with dark lines.

Tiger Cowry *Cypraea tigris* IP: E. Africa to Hawaii, 150 mm. Similar to *C. pantherina,* areas do not overlap.

Tiger Cowry *Cypraea tigris* (continued) Feeds on sponges, bryozoans and algae. Photo on the left - animal with a mantle retracted. Photo on the right - juvenile, "bulla" stage.

Cox's Cowry *Eclogavena coxeni* WP: New Guinea to Solomons, 32 mm. Elongated white shell with brown spots, grey-green mantle with scattered reddish papillae. ©SJ.

Four-Spotted Cowry *Eclogavena quadrimaculata* WCP, 32 mm. Cylindrical shell with four very pale bands and four dark spots near ends. Dark mantle with pinkish spots. Right: © Ria Tan.

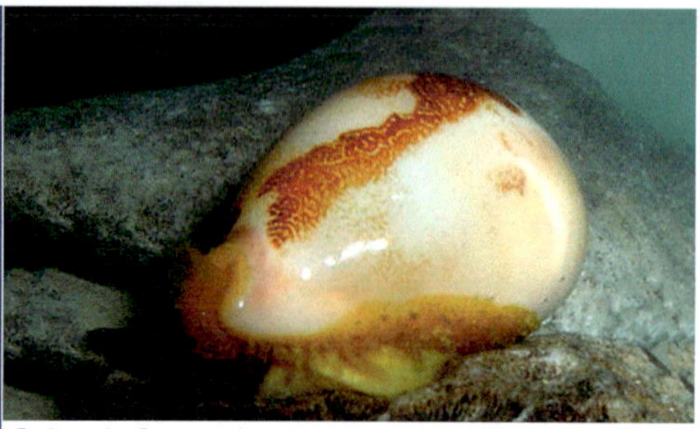

Sakura's Cowry *Austrasiatica sakuraii* WP: Philippines Japan, 60 mm. Depth>60 m. © E.G de Suduiraut.

Rabaul's Cowry *Erronea rabaulensis* WP: N. Guinea, Philippines, 33 mm. © E.G de Suduiraut.

Persian Tanned Cowrie *Erronea adusta persica* IO: Persian Gulf to Malaysia, 51 mm. Dark brown shell with two transverse pale bands, brown mantle with white spots around papillae. © Massimo Scali.

Melanesian Succincta *Erronea adusta melanesiae* WCP: Malaysia to Melanesia, 50 mm. ©SJ.

Five-Banded Cowry *Erronea caurica quinquefasciata* Red Sea, WIO, 59 mm. © Massimo Scali.

Thick Edged Cowry *Erronea caurica* IP: Red Sea and East Africa to WCP (not to Hawaii), 70 mm. Creamy shell with pale-brown transverse bands, white foot with black lines. Mantle and tree-like papillae with granulated texture.

Cylindrical Cowry *Erronea cylindrica* IP: Madagascar to N. Caledonia and Fiji , 54 mm. Cylindrical shell, cream with dark-brown blotches. Left: © Y. Thevenet, right: © Philippe & Guido Poppe www.poppe-images.com

Erroneous Cowry *Erronea errones* IWP: East Africa to West Pacific, 45 mm. Cream to olive-brown shell with faint transverse bands, grey to brownish mantle with white papillae. Right photo: ©JJ.

Onyx Cowry *Erronea onyx* IP: East Africa to Solomons and Palau, 60 mm. Elongate-ovate to pear-shaped shell, golden-brown or even jet black to pale with longitudinal brown stripe. Appears to be a species complex. ©SJ.

Onyx Cowry *Erronea onyx* (cont'd) Mantle dark grey to brown with pinkish papillae, tan tentacles, black siphon. Found in shallow sand and mud areas, feeding on algae and coral polyps. © Ria Tan.

Egg Cowry *Erronea ovum* WCP: from Thailand and Malaysia to Solomons and Palau, 42 mm. Olive-brown shell with faint brown bands, mantle generally grey with brighter and darker areas. © Ria Tan.

Greenish Cowry *Erronea subviridis* WP: Australia, New Guinea to Fiji, 42 mm. Cream-olive, ovate to pear-shaped shell with brown blotches. Grey-greenish mantle with tree-like papillae. Left: © David Massemin.

Pear-Shaped Cowry *Erronea pyriformis* IWP, 45 mm. Pale shell with faint brown bands. © Ria Tan.

Children's Cowry *Ipsa childreni* WCP, 32 mm. Shell with transverse ridges. Nocturnal. ©JJ.

Map Cowry *Leporicypraea mappa* IP, 10 cm. Shell globose with highly variable wavy brown pattern. Translucent mantle with thin tapering papillae. Nocturnal, found in caves and overhangs. Appears to be a species complex.

Isabel's Cowry *Luria isabella* IP: Red Sea, East Africa to Hawaii, 54 mm. Cylindrical beige to pale reddish-brown shell, translucent-black mantle, dark grey foot, siphon and tentacles. Nocturnal, feeds on sponges, algae, corals.

Carnelian Cowrie *Lyncina carneola* IP: Red Sea to Hawaii, 66 mm. Shell pale orange-brown with pale transverse bands, mottled reddish brown to grey mantle with white papillae.

Carnelian Cowrie *Lyncina carneola* (cont'd) Photo on the left: with eggs, photo on the right - juvenile (bulla stage, 2 cm).

Checkerboard Cowrie *Luria tessellata* WCP, South East Asia to Hawaii, 58 mm. Easily recognizable animal, shell beige with pale transverse bands and dark brown blotches, translucent mantle.

Leviathan Cowry *Lyncina leviathan* IP: Red Sea to Hawaii, 100 mm. Similar in appearance to *L. carneola,* but larger, the radula (feeding structure like ribbon with teeth) is different. Papillae with numerous branches.

Lynx Cowrie *Lyncina lynx* IP: Red Sea to Hawaii, 85 mm. Shell white to tan with brown spots, adults with bluish cast over the dorsum. Translucent mottled mantle, branching papillae. Common, found under shallow rocks.

Neighboring Cowry *Lyncina propinqua* IP: Red Sea to Hawaii and Melanesia, 52 mm. Similar to *L. carneola,* but shell is wider and generally smaller. DNA analysis confirmed the validity of this species. ©JJ.

Schilder's Cowrie *Lyncina schilderorum* CP, Hawaii 45 mm. Dorsum with distinct central area of beige and orange-brown bands. Mottled mantle with opaque white areas. ©SJ.

Groove Toothed Cowry *Lyncina sulcidentata* CP: Hawaii, 60 mm. Similar in appearance to *L. schilderorum*, but tends to be bigger, dorsal bands are less distinct. ©SJ.

Ventral Cowry *Lyncina ventriculus* WCP: Indonesia and Philippines to Polynesia, 76 mm. Shell golden-brown with paler area mid-dorsally, darker area with fine transverse stripes. Mottled dark-grey mantle, short papillae.

Pacific Deer Cowry *Lyncina vitellus* IP: Red Sea, East Africa to Hawaii, 100 mm. Common and widespread in the region. Shell from pale to dark-brown, with white spots of different size. Photo on the right - juvenile, ©JJ.

Arabian Cowry *Mauritia arabica* IP: East Africa to Tahiti, 75 mm. Shell large, ovate to pear-shaped, chestnut-brown, with longitudinal brown lines, interrupted by pale spots. Mantle translucent.

Arabian Cowry *Mauritia arabica* (cont'd) Numerous elongated, conical papillae are translucent to tan.

Gray's Arabica Cowry *Mauritia grayana* WIO: Red Sea to Tanzania and Pakistan. © Irina Sevriuk

Depressed Cowry *Mauritia depressa* IP: Madagascar and Reunion to Polynesia, 88 mm. Shell with widened lateral margins, brown with white spots. Mantle translucent. Found in the shallow wave-washed areas. ©SJ.

Dog-Rose Cowry *Mauritia eglantina* IP: Madagascar to Polynesia, 95 mm. Similar to *M. arabica*, but tends to be more cylindrical, less oval in shape. ©SJ.

Blotched Cowry *Mauritia maculifera* PO, 89 mm. Similar in appearance to *M. depressa*, but less flattened and often demonstrates a dark blotch on the columellar side of the base in adults. Mantle translucent.

Chocolate Cowry *Mauritia mauritiana* IP: East Africa to Hawaii, 130 mm. Beautiful dark shell with golden-brown round spots. Semi-translucent mantle with thin papillae. Found in the shallow wave-washed areas. ©SJ.

Jester Cowry *Mauritia scurra* IP: E. Africa-Hawaii, 60 mm. Long cylindrical shell, translucent mantle. ©JJ.

Lister's Cowry *Melicerona listeri* Shell cream with brown dots and faint bands. IWP, 28 mm. ©JJ.

Ring Cowry *Monetaria annulus* IP: Red Sea, East Africa to Hawaii, 50 mm. Shell white, with two yellow stripes along the upper side. Was used as as shell money and in ornaments. Found in large numbers in intertidal rocks.

Ring Cowry *Monetaria annulus* (cont'd) Photo on the right demonstrates difference between *M. annulus* and *M. moneta* (lower right). Left: © Yves Thevenet (on eggs).

Money Cowry *Monetaria moneta* IP, 46 mm. Moderately small shell with with widened lateral margins and swollen dorsal side. Mantle white with vertical grey lines.

Snake-Head Cowry *Monetaria caputserpentis* IP: East Africa to Hawaii and Galapagos, 45 mm. Reddish-brown shell with round white spots on the top of the dorsum. Mottled mantle with short papillae. ©SJ (left), ©JJ (right).

Beck's Cowry *Naria beckii* IP: East Africa to Polynesia, 20 mm. Beige to orange shell with red-brown and white spots, translucent mantle with very long tree-like papillae. Photos: ©SJ (left), ©JJ (right).

Boivin's Cowry *Naria boivinii* IWP: India to Philippines and Japan, 40 mm. Shell beige-grey with bluish rounded spots. Mottled greenish mantle with very long tree-like papillae.

Fawn Cowry *Naria nebrites* IWP: Red Sea and Mozambique to Sri Lanka and Singapore, 50 mm. Shell orange to brown with white dots, mottled mantle with tree-like papillae. © Irina Sevriuk.

Ivory Cowry *Naria eburnea* WP: New Guinea to New Caledonia and Fiji, 59 mm. Totally white shell, olive-brown mantle with tree-like papillae. Right: © Yves Thevenet.

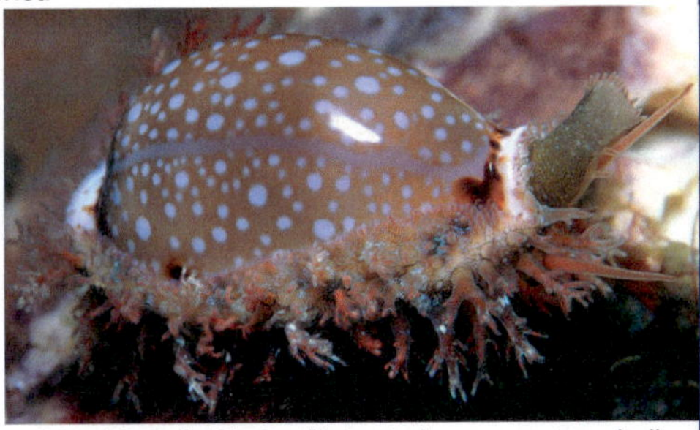

Lined-Lip Cowry *Naria labrolineata* IP: East Africa, Indonesia to Hawaii, 31 mm. Beige to brown-orange shell with rounded white spots of different size, dorsum with pale longitudinal line.

Lined-Lip Cowry *Naria labrolineata* (cont'd) Mottled brown-olive mantle with long reddish-green papillae.

Separated Cowry *Naria cernica* IP: E. Africa, Maldives to Australia and Hawaii, 46 mm. © Brian R. Mayes.

Eroded Cowry *Naria erosa* IP: East Africa to Marshall Islands, 75 mm. Pale-brown shell with small white spots and dark brown spots on the extremities. Nocturnal, hides under rocks during the day.

Eroded Cowry *Naria erosa* (cont'd) Juveniles paler, without white spots.

Strawberry Cowry *Naria helvola* IP, 36 mm. Mantle densely covered with olive to pink papillae.

Strawberry Cowry *Naria helvola* (cont'd) Shell smooth, elongated, red-brown with darker areas and small white spots. Photo on the right: on eggs.

Military Cowry *Naria miliaris* IWP: Madagascar, Tanzania to Indonesia and Philippines, 60 mm. Shell cream to brown or yellowish, with small white spots. Mottled brown mantle with tree-like papillae.

Porous Cowry *Naria poraria* IP: East Africa to Polynesia, 30 mm. Brown-purplish shell with small white spots, partly brown-bordered. Mantle with tapering red-purple papillae. ©SJ.

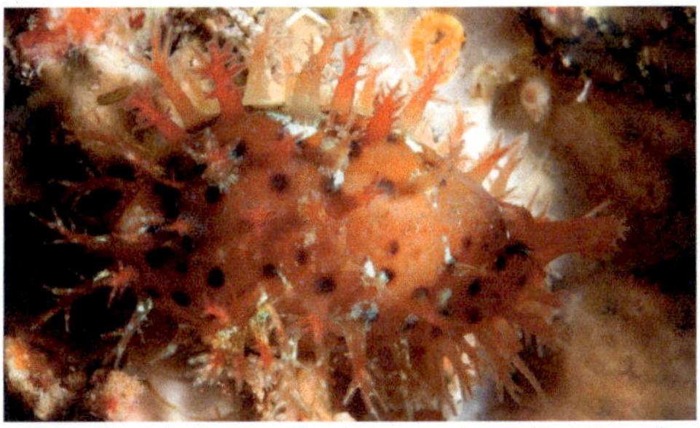

MacAndrew's Cowry *Naria macandrewi* Red Sea, E. Africa, Oman, 27 mm. Shell with round brown spots.

Thrush Cowry *Naria turdus* Red Sea, WIO, 62 mm. Intertidal, white with brown dots. © Ravi Vadher.

Nucleus Cowry *Nucleolaria nucleus* IP: Red Sea, East Africa to Fiji and Marshall Islands, 37 mm. Unusual granulose shell, mantle with very long papillae. Photo on the right: adult (©JJ), left - juvenile.

Granulated Cowry *Nucleolaria granulata* CP: Hawaii and Marquesas Islands, 49 mm. Flattened granulose shell with thin transverse ribs. Mottled mantle with very long papillae. Nocturnal, hides under rocks during the day. ©SJ.

Chinese Cowry *Ovatipsa chinensis* IP: E. Africa to Central Pacific, 52 mm. Shell red-brown with white spots mid-dorsally, with purple spots on the edges. Mantle bright red. Photo on the right (on eggs): ©JJ.

Rashleigh's Cowry *Ovatipsa rashleighana* Hawaii, 29 mm. White shell with red-brown speckles and wide band in the middle. Red mantle with numerous hair-like papillae in addition to larger simple white ones. ©SJ.

Asellus Cowry *Palmadusta asellus* IP: East Africa to Marshall Islands, 24 mm. White shell with three wide brown transversal bands on the upper surface. Nocturnal, hides under shallow rocks during the day.

Clandestine Cowry *Palmadusta clandestina* IP: Red Sea and East Africa to Marshall Islands, 26 mm. Shell with fine transversal orange lines, mantle grey-olive to orange-brown. Prefers shallow lagoons, nocturnal.

Clandestine Cowry *Palmadusta clandestina* (continued) Like some other cowries, secretes sulphuric acid when disturbed. ©JJ.

Contaminated Cowry *Palmadusta contaminata* IP: East Africa to Polynesia, Marquises, 20 mm. Beige-bluish shell with wide brown transverse bands and orange brown spots - often one big and several smaller ones.

Contaminated Cowry *Palmadusta contaminata* (continued) Mantle orange-brown to dark grey, with opaque white spots and translucent-white papillae. Photo on the right: on eggs.

Humphrey's Cowry *Palmadusta humphreyii* WCP: Australia to Melanesia, 27 mm. White with numerous brown spots, orange edges. Red mantle with small tapering papillae. ©SJ.

Freckled Cowry *Palmadusta lentiginosa* IO: Red Sea to India, 40 mm. White with numerous brown spots, aggregated in the wide transverse bands. Orange granulated mantle with white papillae. © Massimo Scali.

Yellow Cowry *Palmadusta lutea lutea* WP: Indonesia to Marshall Islands, 27 mm. Shell orange to grey-brown with two transverse white bands and small brown spots. Orange mantle with short papillae.

Yellow Cowry *Palmadusta lutea johnsonorum* WCP: Marshall Islands, Guam, 17 mm. Differs from *P. lutea lutea* (above) in a more red color and heavier spotting over the shell. ©SJ.

Saul's Cowry *Palmadusta saulae* IWP: Andaman Islands to New Caledonia, 35 mm. Pale blue shell with faint transverse bands and brown spots. Bright orange mantle with thin papillae. © Ria Tan.

Zigzag Cowry *Palmadusta ziczac* IP: Red Sea, E. Africa to Polynesia, 26 mm. Light beige or tan pyriform shell, 3 transversal white bands with a zigzag pattern. Orange to red semi-translucent mantle with tree-like papillae.

Hungerford's Cowry *Paradusta hungerfordi* WP: Japan to Australia, 48 mm. © E.G de Suduiraut.

Drop Covered Cowry *Perisserosa guttata* IWP: India to Philippines, 87 mm. © E.G de Suduiraut.

Fringed Cowry *Purpuradusta fimbriata* IP: Red Sea and East Africa to Hawaii and Marquesas, 21 mm. Small almost cylindrical shell, white with reddish brown dots and wide transverse pale-brown bands. Mantle red-orange.

Graceful Cowry *Purpuradusta gracilis* IP: Red Sea and East Africa to Fiji and Japan, introduced to Mediterranean, 30 mm. White or pale-bluish with numerous small brown spots, grainy mantle and tapering papillae.

Small-Toothed Cowry *Purpuradusta minoridens* IWP Sri Lanka to Marshall Isl, 13 mm. Small pale brown-dotted shell with very faint transverse bands, purple mark at anterior end. Red mantle with white papillae.

Mrs Hammond's Cowry *Purpuradusta hammondae* WCP, 20 mm. Pale cream with brown speckles. ©SJ.

Tiny-Teeth Cowry *Purpuradusta microdon* IP: Red Sea to Marshall Islands, 18 mm. ©JJ.

Tiny-Teeth Cowry *Purpuradusta microdon* (continued) Pale tan or cream shell with small brown spots and faint transverse bands. Similar to *P. fimbriata* and *P. minoridens,* but less cylindrical, more rounded.

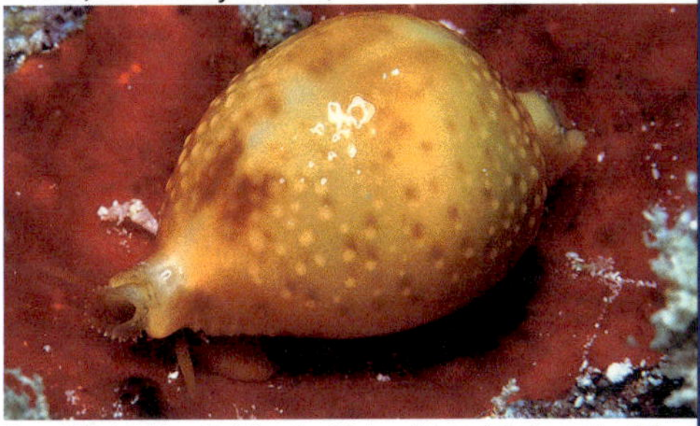

Orange Chick Pea Cowry *Pustularia bistrinotata* IP: Red Sea, East Africa to Marshall Islands and Polynesia, 25 mm. Orange-brown bumpy shell with extremely produced ends, translucent mantle. Photos: (left, right): ©SJ, ©JJ.

Chick-Pea Cowry *Pustularia cicercula* IP: East Africa to Central Pacific, 25 mm. Shell with white to creamcolored bumpy dorsum, mottled grey mantle. Nocturnal, hides deep in rubble during the day. ©JJ.

Globular Chick-Pea Cowry *Pustularia globulus* IP: Red Sea to Marshall Isl, absent in Hawaii. Light orange to dark brown smooth shell with dark spots, translucent mantle. Photos: (left, right): ©JJ, ©SJ.

Margarita Chick Pea Cowry *Pustularia margarita* IP: Red Sea, Mauritius to Philippines and Marshalls, 21 mm. Similar in coloration to *P. cicercula*, but shell is smooth, with distinct orange-brown spots. ©JJ.

Maui Chick-Pea Cowry *Pustularia mauiensis* CP: Hawaii, 16 mm. Smooth white shell with brown spots. Mottled grey mantle. Nocturnal, hides under rocks and rubble during the day. ©SJ.

Slug-Like Cowry *Staphylaea limacina* East Africa to Marshall Islands (absent in Hawaii), 39 mm. Similar in appearance to *S. staphylaea* but has coarse sparse warts rather than the finely pustulose dorsum.

Slug-Like Cowry *Staphylaea limacina* (cont'd) Orange to red-brown mantle with very long finger-like bluntended papillae. Nocturnal, hides under shallow rocks during the day.

Little Spotted Cowry *Staphylaea semiplota* CP: Hawaii, 37 mm. Cream to dark-brown shell with small white spots. Dark mantle with dense tapering papillae. ©SJ.

Pustulose Cowry *Staphylaea staphylaea* IP: Red Sea to Marshall Islands, 29 mm. Bluish or red-brown finely pustulose shell with orange ends. Nocturnal, hides under rocks during the day. Photo on the left ©JJ.

Pustulose Cowry *Staphylaea staphylaea* (cont'd) Dark-brown mantle blends in well with their typical food source, a black encrusting sponge. Unlike *S. limacina*, papillae with pointed, often white apex. ©SJ.

Martin's Cowry *Ransoniella martini* WCP: Indonesia, Philippines to N. Caledonia and Marshall Islands, 21 mm. Elongated, like immature shell with brown spots and faint transverse bands. Mantle with white papillae. ©JJ.

Punctate Cowry *Ransoniella punctata* IP: Red Sea, East Africa, to Marshall Islands, 22 mm. Elongated white shell with dark-brown round spots. Red mantle with with short white papillae.

Burgess' Cowry *Talostolida latior* CP: Hawaii, 47 mm. Wide shell with black marginal spotting, brown speckles and faint transverse bands. Numerous small papillae, black at the base with branching white tips. ©SJ.

Pellucent Cowry *Talostolida pellucens* IP: East Africa to Pacific Mexico, 44 mm. White to yellowish shell with brown speckles and transverse bands. Red mantle and foot. Often found under shallow rocks during the day.

Pellucent Cowry *Talostolida pellucens* (continued) Similar in appearance to *T. teres*. These two genetically distinct species are extremely hard to distinguish by photo. Photo on the left - animal on eggs (©SJ).

Tapering Cowry *Talostolida teres* IP: Red Sea to Galapagos, 47 mm. Shell yellowish white with wide transverse bands of darker brown in middle region. Found under shallow rocks, feeds on red sponges.

Tapering Cowry *Talostolida teres* (continued) Photo on the left: juvenile. Right - *Talostolida teres* var. *alveolus*. Like *T. pellucens, T. teres* will autotomize the rear portion of its foot when disturbed.

Mole Cowry *Talparia talpa* IP: Red Sea to Polynesia, 113 mm. Yellowish shell with red-brown transverse bands. Foot and mantle dark grey with white dots, unusual rounded papillae. Left: © Irina Sevriuk, right: ©JJ.

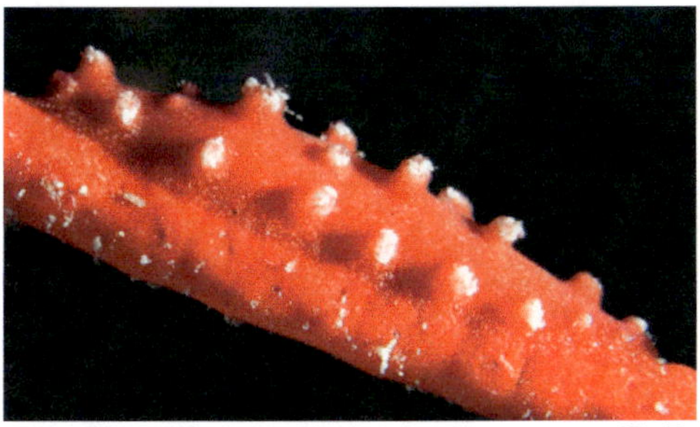

White-Tufted Spindle Shell *Aclyvolva coarctata* IP: East Africa to Marshall Islands, 30 mm. *Hiatavolva rugosa* a junior synonym. Lives on red to yellow-brown whip-like gorgonians in the genus *Ellisella*.

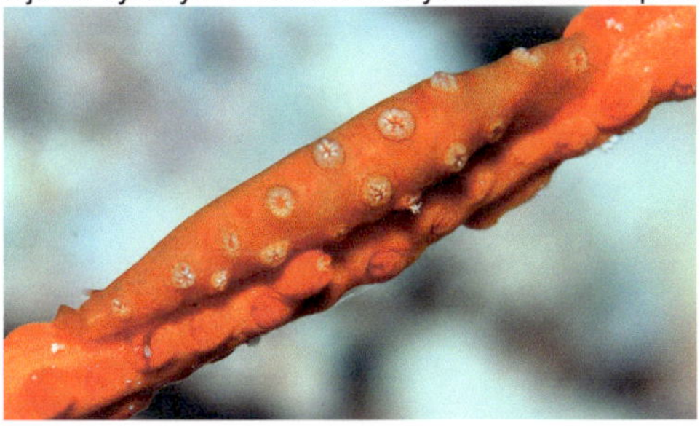

White-Tufted Spindle Shell *Aclyvolva coarctata* (continued) Mantle coloration matches the color of the host gorgonian. Compound tuft-like retractable papillae mimic the polyps of the host. ©JJ.

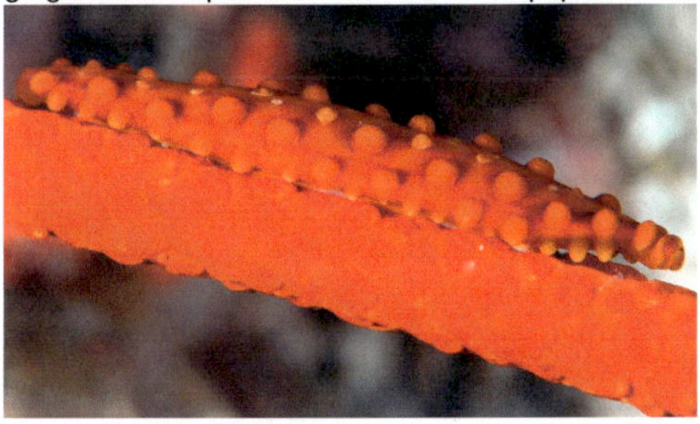

Lance Spindle Shell *Aclyvolva lanceolata* IP: Red Sea and East Africa to Guam and Marshall Islands, 35 mm. Elongated shell, mantle with rounded compound papillae. Papillae mimic the polyps of the host gorgonian.

Lance Spindle Shell *Aclyvolva lanceolata* (continued) Color morphs, close to the description of *A. lamyi*, a synonym of *A. lanceolata* now. ©JJ, ©SJ (left, right).

Lance Spindle Shell *Aclyvolva lanceolata* (continued). Color morph, close to the description of *A. nicolamassierae*, a synonym of *A. lanceolata* now. Left (Red Sea): © Sven Kahlbrock, right (Marshall Isl.): ©JJ.

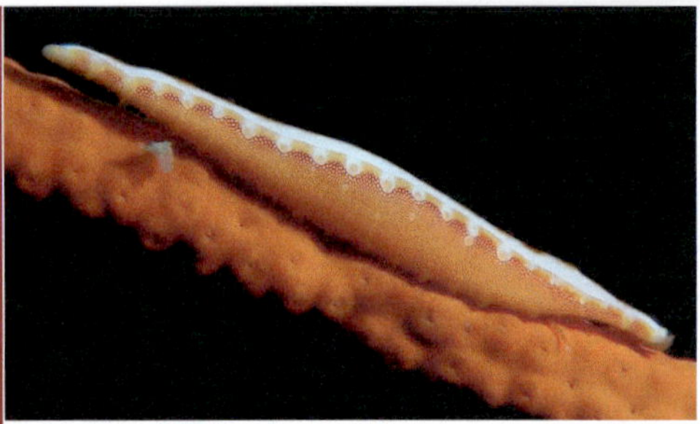

Graceful Spindle Shell *Calcarovula gracillima* IWP: Japan, South Africa, Red Sea, Solomons, 30 mm. Found on *Dichotella* sea whips. Red or brown mantle with scalloped border and white edge. © Johji Nishio (Japan).

Graceful Spindle Shell *Calcarovula gracillima* (cont'd) © Sven Kahlbrock (Red Sea).

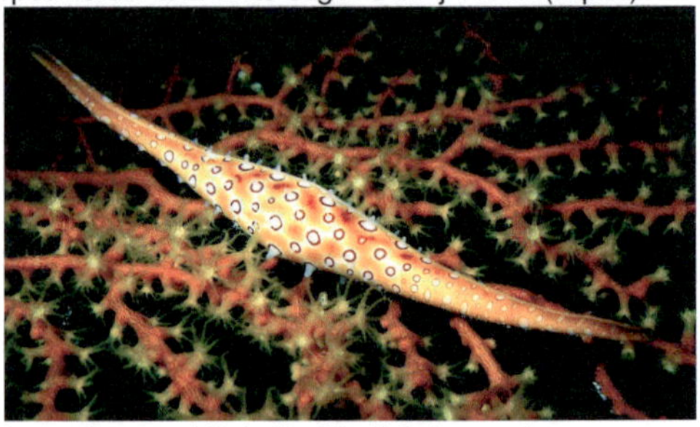

Long-Snouted Spindle Shell *Calcarovula longirostrata* IP, 70 mm. © Sven Kahlbrock (Red Sea).

Long-Snouted Spindle Shell *Calcarovula longirostrata* (con-d) Described from the Adriatic Sea, reported from the Red Sea, E. Africa and Japan. Found on *Dichotella* sea whips, *Astrogorgia* sea fans. © J. Nishio. (Japan).

Depressed Spindle Shell *Hiatavolva depressa* WP, 31 mm. Found on *Alertigorgia* bushy corals.

Shingo's Shell *Kuroshiovolva shingoi* IWP, 30 mm. On *Plumarella cristata*, 120 m+. © E.G de Suduiraut.

Wakayama's Spindle Shell *Kurodavolva wakayamaensis* IP: Reunion to Japan, 76 mm. © Johji Nishio.

Malaitan Egg Shell *Naviculavolva malaita* WCP, Indonesia to Marshalls, 12 mm, on *Rumphella* fans. ©JJ.

Kurz' Egg Shell *Naviculavolva kurziana* WCP: Indonesia to Marshalls, 17 mm. Feeds on *Rumphella gorgonian*. Species delimitation between *N. kurziana* and *N. malaita* (above) needs to be confirmed with genetic data. ©JJ.

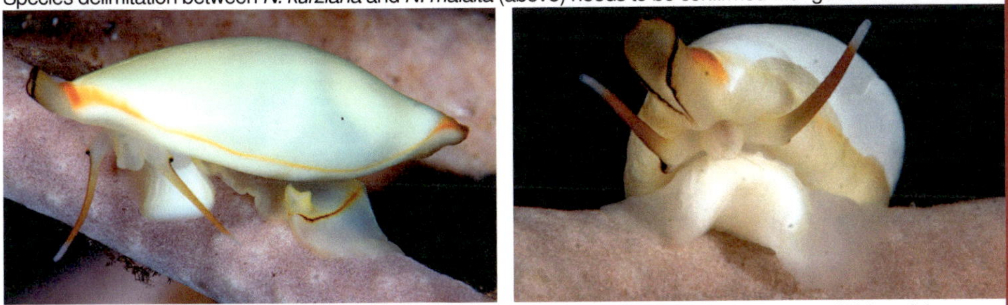

Canoe Egg Shell *Naviculavolva deflexa* WP: Philippines, Indonesia, Australia, PNG, 22 mm. White, pink or grey animal with translucent mantle, shell usually with orange suture line. Feeds on *Rumphella* coral.

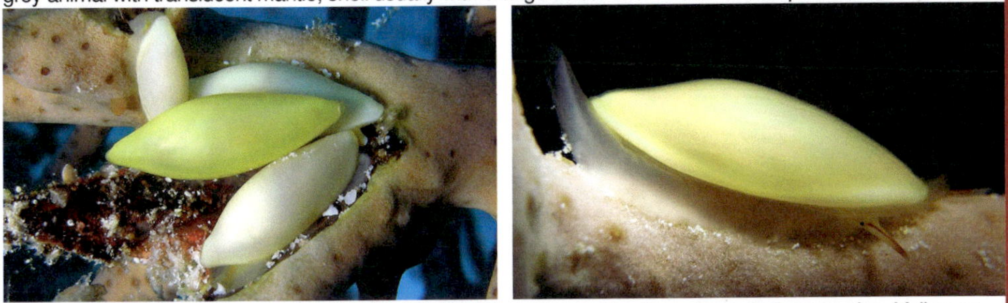

Debelius' Egg Shell *Naviculavolva debelius* IO: Red Sea, 15 mm. Feeds on *Rumphella* gorgonian. Yellow or white shell, translucent-whitish mantle, white tentacles with red-brown eye-stripe. © Sven Kahlbrock.

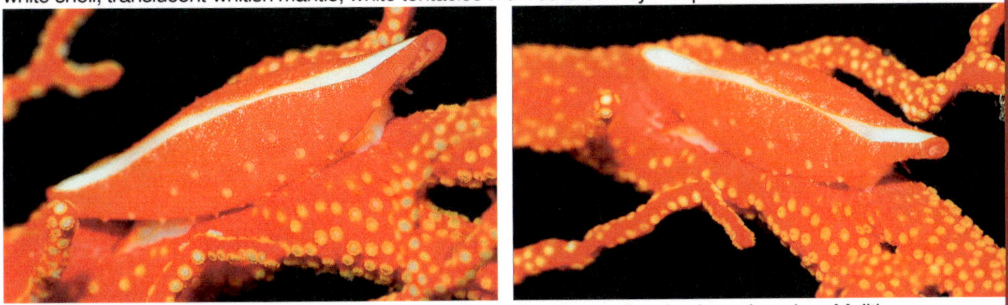

Angas' Egg Shell *Pellasimnia angasi* WP: Japan, Australia, 38 mm. Sea fan form, found on *Melithaea* sea fans. Black coral form (below) appears to be the same species. © Brian R. Mayes.

Angas' Egg Shell *Pellasimnia angasi* Black coral form, feeds on *Myriopathes* black corals. Morphologically identical to Sea fan form. Mottled mantle with tree-like papillae (extended) © Sue Churchill & Rogan Draper.

Annabel's Spindle Shell *Pellasimnia annabelae* WP to Marshall Islands, 35 mm, feeds on *Annella* gorgonians. Morphologically close to *P. brunneiterma* (below), species delimitation was not confirmed with genetic data.

Brown-Tipped Spindle Shell *Pellasimnia brunneiterma*, IWP, 35 mm. On *Annella* gorgonians. Left: Red Sea, © Sven Kahlbrock with smaller *P. annabelae*, possibly the same species. Right: © Rokus Groeneveld.

Orange-Tipped Spindle Shell *Pellasimnia improcera* WP: Korea to Solomons, 35 mm. Form on *Myriopathes* black corals. Shell with orange tips, mantle with red or orange lines and tapering white papillae. Right: ©JJ.

Pellasimnia improcera (cont'd) Form on on *Melithaea* sea fans. © Brian R. Mayes.

Golden Spindle Shell *Pellasimnia* sp. SP: Australia, 25 mm, on *Iciligorgia* sp. © Sue Churchill & Rogan Draper.

Barbier's Spindle Shell *Phenacovolva barbieri* WP: Japan to Singapore, Malaysia, Brunei, 43 mm. Feeds on gorgonians. Transparent mantle with rounded black spots and yellow tapering papillae. © Brian R. Mayes.

Spindle Cowries on this page can be tentatively identified as follows: *Phenacovolva gracilis*: "a", *Phenacovolva pseudogracilis*: "b"-"h", *Phenacovolva subreflexa* "j"-"i". They are the species with Western Pacific distribution, feeding on black corals, shell size up to 35-38 mm.

Species delimitation in this group needs to be reconsidered with genetic data. It is highly likely that they are the morphs of the same species. Until a revision of this group of species is published it seems optimal to follow Draper & Churchill 2014 and "consider them all the same species, using the original name *Phenacovolva gracilis*". Photo e,f,j,i - ©JJ.

Short-Snouted Spindle Shell *Phenacovolva brevirostris* IWP, 40 mm, feeds on gorgonians. High profile shell with more or less distinct white line across the shoulder, translucent mantle.

Fused Spindle Shell *Phenacovolva fusula* E. Africa to Japan, 29 mm. Hosted on *Acanthogorgia* sea fan. The shell has 3 wide dorsal bands, short terminals tipped in orange/brown. © Brian R. Mayes.

Iredale's Spindle Shell *Phenacovolva nectarea* IWP, 65 mm. Translucent pinkish mantle with red spots, large conical white and yellow papillae with a rounded white areas around the base. Photo: © S. Churchill & R. Draper.

Iredale's Spindle Shell *Phenacovolva nectarea* (continued) Feeds on *Euplexaura* sea fans. Close to *P. rosea* but shell is more robust. © Ria Tan.

Poppe's Spindle Shell *Phenacovolva poppei* IWP: East Africa to Indonesia and Philippines, 38 mm. Feeds on black corals. High profile shell, white mantle with pattern of dark red lines, long tapering papillae.

Rosy Spindle Shell *Phenacovolva rosea* IP: East Africa to Central Pacific, 65 mm. The most widespread and common species of Ovulidae. Translucent mantle with red-brown spots of varying size. Shell with more or less distinct white line across the shoulder. Large reddish, yellow or white papillae are accompanied by a pattern of much smaller white conical papillae.

Feeds on gorgonians, roosting on a record number of host species.

Photos "e"-"j" - ©JJ, "i" - ©SJ.

Pink Spindle Shell *Phenacovolva* sp. WP: Philippines, 40 mm. Undescribed species with branching papillae.

Okinava Spindle Shell *Phenacovolva*? sp. Japan, 20 mm. Shell with upturned ends. © Shawn Miller. IT

Schmid's Spindle Shell *Phenacovolva schmidi* WP: Andaman Sea to Philippines and Australia, 53 mm. White mantle with scattered red spots and short papillae, shell with yellowish tips.

Little Spindle Shell *Phenacovolva parvita* WP: Japan, Philippines, Indonesia, 25 mm. Mantle with fine reticu-lation, red upper part. On *Paratelesto* gorgonians. © Philippe&Guido Poppe www.poppe-images.com.

Curved Spindle Shell *Phenacovolva recurva* IP: South Africa, Japan to Indonesia. Orange mantle with reticulated pattern of white or yellow papillae. On *Paratelesto* gorgonians. © Rokus Groeneveld.

Rehder's Spindle Shell *Phenacovolva rehderi* WP, Japan to Brunei, 44 mm. Hosted on *Myriopathes* sp. black coral. Easy to tell apart from *P. gracilis* complex (above) due to plain white or yellowish shell. © Brian R. Mayes.

Robust Spindle Shell *Phenacovolva* sp. WP: Philippines, Malaysia, Brunnei, 38 mm. Hosted on *Dendronephthya* sp. soft coral. Appears to be undescribed. White mantle with small dark spots. © Brian R. Mayes.

Elongated Egg Shell *Volva volva* IWP: East Africa to Japan, Indonesia, Philippines, Australia, 186 mm. Shell with long slender terminals. Mantle with brown patches and long, brown-tipped papillae.

Elongated Egg Shell *Volva volva* (continued) found in muck areas. Feeds on octocorals, including sea pens. Animal on the photo was found finishing up *Dendronephthya* coral and laid eggs in the remnants. ©JJ.

Tall Egg Shell *Carpiscula procera* WP: Taiwan, Brunei, Papua New Guinea, 13 mm. Feeds on sea fans. Strongly transversely striate shell, color mimic the color of host gorgonian. Mantle with brown dots. © M. Kiser.

Blistered Egg Shell *Carpiscula bullata* WP, 15 mm. © Philippe & Guido Poppe www.poppe-images.com

Elegant Egg Shell *Diminovula concinna* IWP: E. Africa to Philippines and Indonesia, 18 mm. ©JJ.

Elegant Egg Shell *Diminovula concinna* (cont'd) Found on *Capnella* and *Paralemnalia* octocorals. White mantle with polygonal orange islands. Fine transverse lines on the foot.

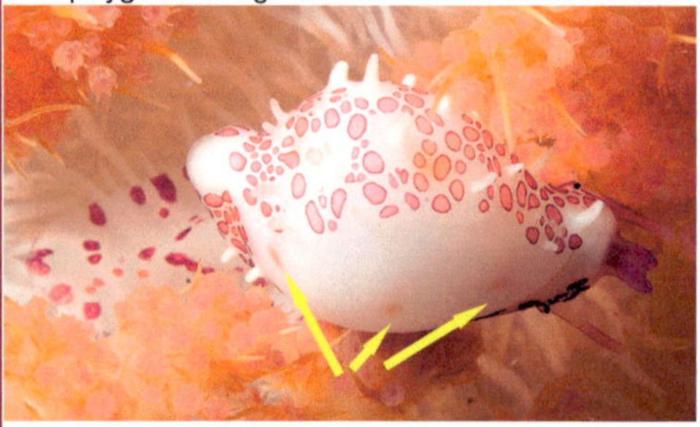
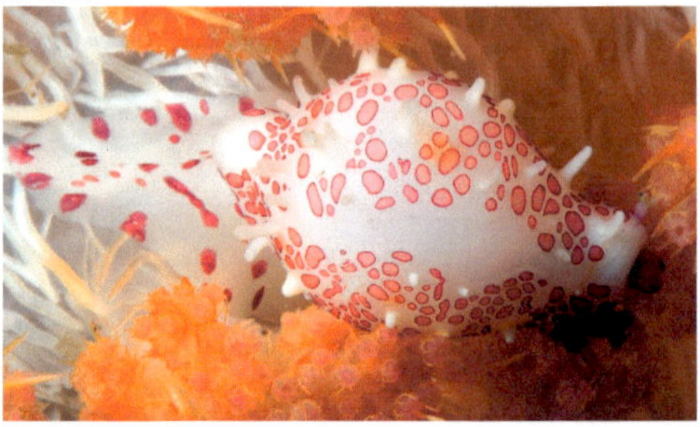

Rose Egg Shell *Diminovula coroniola* WP, 15 mm. Found on *Dendronephthya* corals. Translucent mantle with polygonal pink islands. Shell with dark spots in a rows (highlighted by arrows). Foot with blurry transverse.

White Egg Shell *Diminovula alabaster* IWP: 15 mm. *D. alabaster, coroniola, aurantiomacula, punctata, kosugei,* whitworthi are hard to tell apart in the field. Treated by Draper & Churchill 2014 as a single variable species. © R. Tan.

Yellow-Spotted Egg Shell *Diminovula aurantiomacula* WP: Australia to Japan, 11 mm. Shell with yellow/ brownish spots. Feeds on *Dendronephthya* soft corals. Left: © Marty Kiser, right: ©SJ.

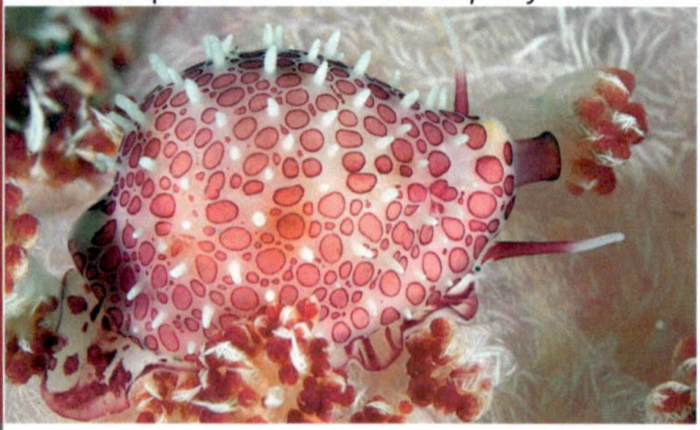

Punctate Egg Shell *Diminovula punctata* IWP: East Africa to Japan and Australia, 15 mm. Shell with dark spots in a rows. Feeds on *Dendronephthya* soft corals. Look comment to *D. alabaster.* © Johji Nishio.

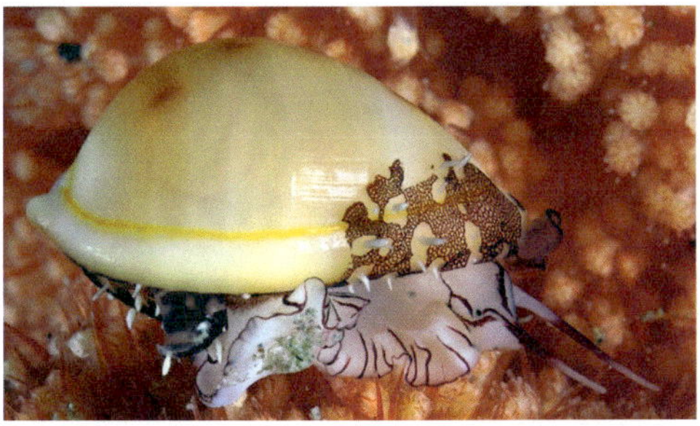

Ridged Egg Shell *Diminovula culmen* IWP, 18 mm. One of the most common and variable species of Ovulidae in the region. White mantle with areas of red reticulations. White papillae with rounded white base.

Ridged Egg Shell *Diminovula culmen* (continued) Feeds on *Dendronephthya* soft corals. Color morph with mantle, mostly covered by red reticulation. © Philippe & Guido Poppe www.poppe-images.com

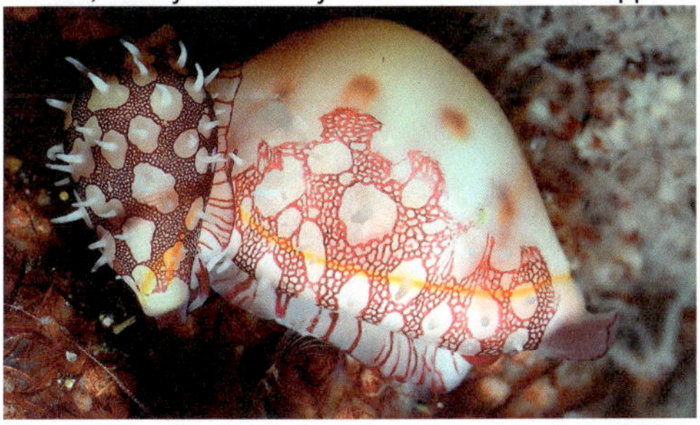

Ridged Egg Shell *Diminovula culmen* (continued) Note brown spots spots and suture line on the shell. Photo on the left: two *D. culmen*, smaller one possibly male (©JJ).

Nielsen's Egg Shell *Diminovula nielseni* WP: Australia to Japan, 11 mm. © Sue Churchill & Rogan Draper.

Fainzilberi's Egg Shell *Diminovula fainzilberi* IO: Red Sea, 8 mm. © Sven Kahlbrock.

Incised Egg Shell *Diminovula incisa* WP: Japan, Philippines, Brunei, 9 mm. Shell with orange blotches, mantle with similar pattern, on *Scleronephthya* sp. soft coral. Photo: (left, right): © E.G de Suduiraut, © Brian R. Mayes.

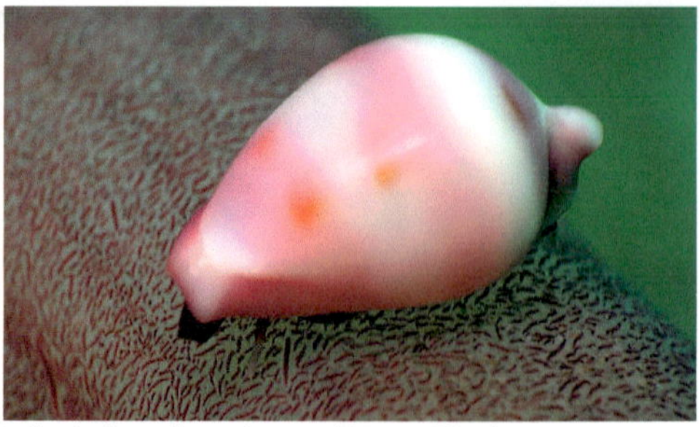

Whitworth's Egg Shell *Diminovula whitworthi* WP: Australia, Indonesia to Japan, 19 mm. Feeds on *Dendronephthya* soft corals. Mantle with red blotches and tapering white papillae. © Brian R. Mayes.

Margarita Egg Shell *Diminovula margarita* WP: Japan to Australia, 13 mm. Translucent mantle with pink rounded spots with dark red outline. Shell with orange suture line, visible at the ends.

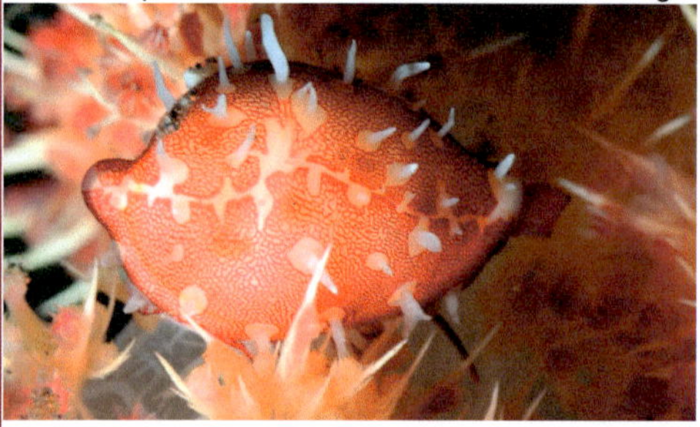

Marked Egg Shell *Diminovula stigm*a WP, 16 mm. Feeds on *Dendronephthya* soft corals. Globular shell with curved terminals. Mantle with red reticulation, tapering white papillae. ©SJ.

Whitworth's Egg Shell *Diminovula whitworthi* WP: Australia, Indonesia to Japan, 19 mm. Feeds on *Dendronephthya* soft corals. Mantle with red blotches and tapering white papillae. © Brian R. Mayes.

Cavanagh's Eggs Shell *Globovula cavanaghi* SP: Australia, 22 mm. Feeds on *Dendronephthya* soft corals. Pink mantle with numerous rounded darker spots and pinkish papillae. © Stephen D.A. Smith.

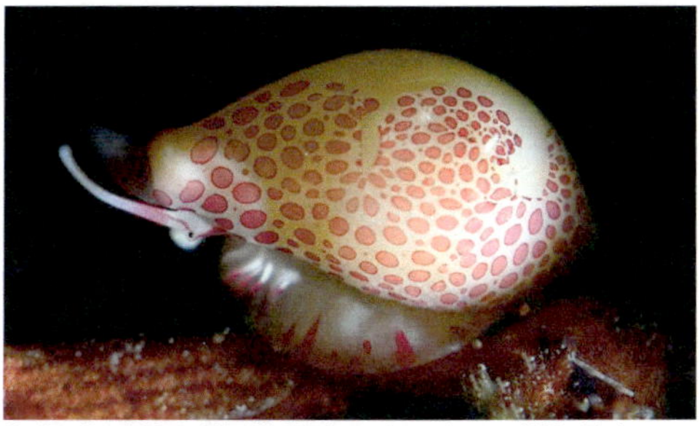

Spherical Egg Shell *Globovula sphaera* WP: Taiwan to Malaysia and Australia, 23 mm. Found on *Dendronephthya* soft corals. White mantle with pink or red spots. Left: © Asyraf Hadee, right: (juvenile) ©JJ.

Enigmatic Egg Shell *Habuprionovolva aenigma* IWP, 10 mm. Feeds on *Dendronephthya* soft corals. Close to *H. basilia* (below) but red lines on the mantle are fine and less distinct. Photo on the right © Brian R. Mayes.

Diadem Egg Shell *Habuprionovolva basilia* IP, 10 mm. Feeds on *Dendronephthya* soft corals. Shell with 3 transverse dark line, visible through the mantle. Right: © Bernd Hoppe.

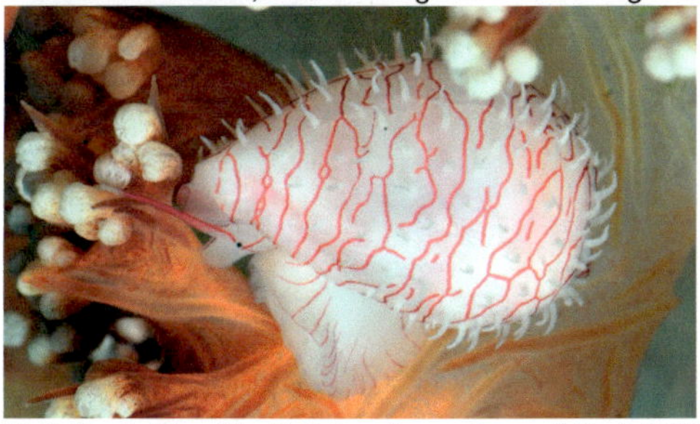

Umbilicate Egg Shell *Habuprionovolva umbilicata* WP: Indonesia, Philippines, China, 10 mm. Uniformly white shell, white mantle with fine red lines and tapering white papillae between. ©JJ.

Ringed Egg Shell *Margovula anulata* WP, 22 mm. Feeds on *Umbellulifera* soft corals. Pink mantle, domed papillae with white areas around. Photos, left: © Marty Kiser, right: © Sue Churchill & Rogan Draper.

Banded Egg Shell *Margovula bimaculata* WP: Brunei to Australia and Philippines, 20 mm. Feeds on *Studeriotes* soft corals. White shell with brown transverse bands, grey mantle mimics the host's polyps. © Brian R. Mayes.

Banded Egg Shell *Margovula bimaculata* (cont'd) Pale form, occurs on paler *Studeriotes* sp. soft corals. © Brian R. Mayes.

Margined Egg Shell *Margovula marginata* IWP, 30 mm. Feeds on *Umbellulifera* soft corals. Yellow-orange mantle with white domed papillae. Photo on the right: © Sue Churchill & Rogan Draper.

Somalian Egg Shell *Margovula somaliensis* IWP: East Africa to Brunei and Philippines, 29 mm. Hosted on *Umbellulifera* soft coral. Globose shell with orange-brown blotches and orange sutural line. © Brian R. Mayes.

Somalian Egg Shell *Margovula somaliensis* (cont'd). Mantle with white domed papillae, surrounded by irregular red-orange pattern. © Brian R. Mayes.

Fruit Egg Shell *Prionovolva brevis* WP: Australia to Philippines, 25 mm. Feeds on *Dendronephthya* corals. Pale shell with wide brownish transverse bands. Mantle with white papillae and red network pattern. Photo: ©JJ.

Fruit Egg Shell *Prionovolva brevis* (continued) Color morph with pronounced red areas around white papillae bases was treated as a separate species before, *P. wilsoniana*. Photo on the right: © Stephen D.A. Smith.

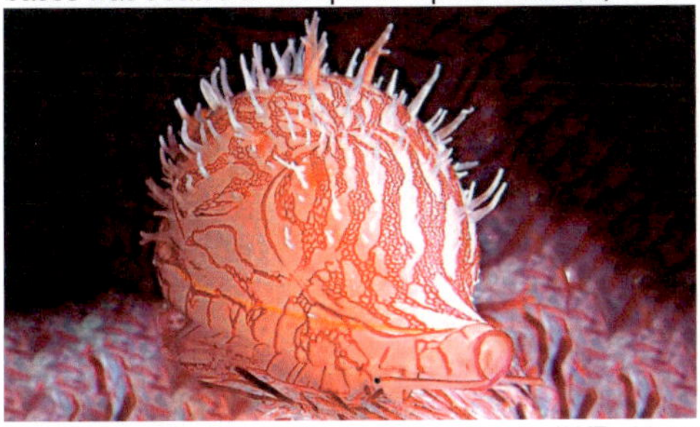
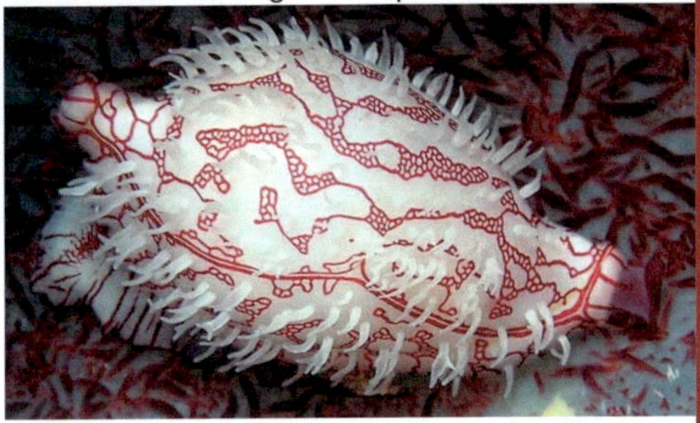

Don Dan's Egg Shell *Serratovolva dondani* WP, 23 mm. Feeds on *Dendronephthya* corals. Mantle with longitudinal red striations. Right: © Sue Churchill & Rogan Draper, left: © Charles E. Rawlings.

Minabe's Egg Shell *Serratovolva minabeensis* IWP: E. Africa, Philippines, Japan, 14 mm. Shell with brown bands, visible through translucent mantle with dark dots. Left: © Johji Nishio, right: © Yui Kawahara.

Nebulous Egg Shell *Testudovolva nebula* WP, 12 mm. Feeds on *Dendronephthya* corals. Shell with blotches that join and swirl. White mantle with pattern of red lines. Right: © B. Hoppe, left: ©JJ.

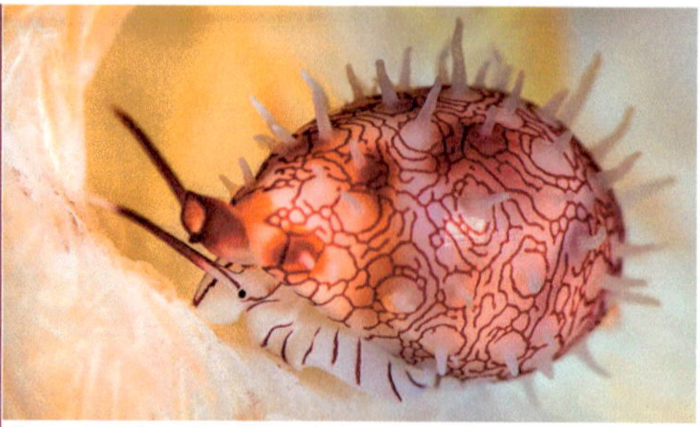

Oriental Egg Shell *Testudovolva orientis* WP: Japan to Indonesia, 10 mm. White shell with small separate brown spots (merged in large blotches in *T. nebula*). Left: © Kelvin Tan HY, right - © Dennis Teck Lee Liew.

Japanese Egg Shell *Testudovolva nipponensis* WP: Japan to Philippines and Brunei, 18 mm. Shell with large blotches that sometimes touch each other. © Brian R. Mayes.

Erica's Egg Shell *Testudovolva ericae* WP, 14 mm. Feeds on *Dendronephthya* corals. Shell is totally white, without spots. White mantle with islands of red lacy pattern. Foot with fine transverse lines.

Erica's Egg Shell *Testudovolva ericae* (cont'd) Rare form with pinkish shell. © Brian R. Mayes.

Jean's Egg Shell *Pseudosimnia jeanae* IWP: S. Africa, Japan to N. Guinea, 21 mm. © E.G de Suduiraut.

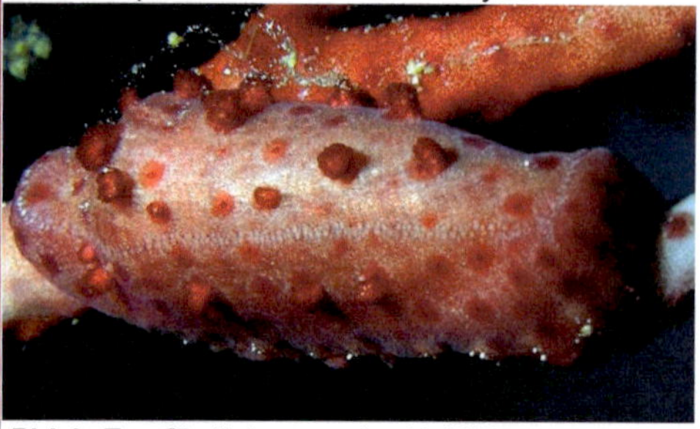

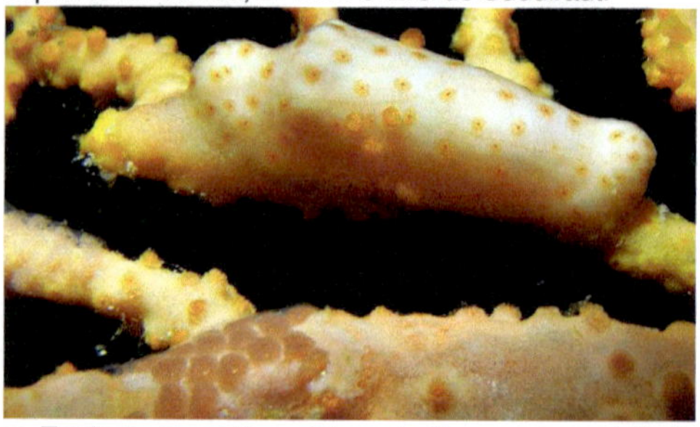

Pirie's Egg Shell *Amonovula piriei* WP, 22 mm. Feeds on *Euplexaura* sea fans. Elongated hump-shaped shell, yellow to red mantle with papillae, matching host's polyps. © Sven Kahlbrock.

Club Egg Shell *Archivolva clava* PO: Papua New Guinea to Japan and Marshall Islands, 17 mm. Found on *Siphonogorgia* octocorals. Brownish shell with hump, more prominent on smaller animals. ©JJ.

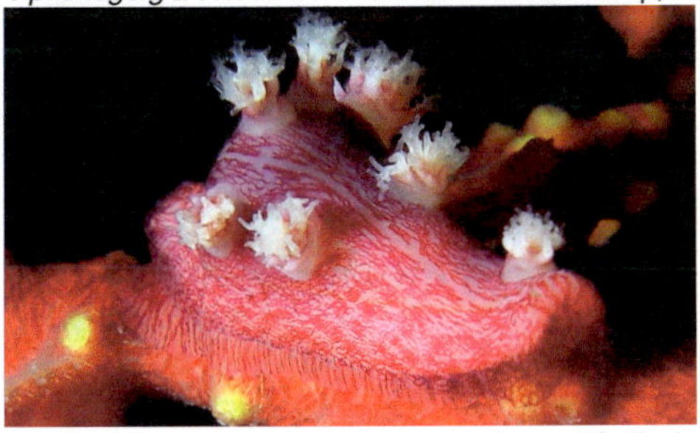

Kahlbrock's Egg Shell *Archivolva kahlbrocki* IO: Red Sea, 15 mm. © Sven Kahlbrock.

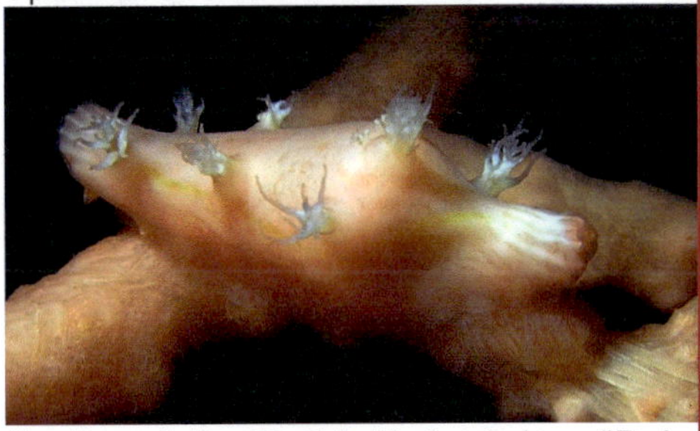

Alex Brown's Egg Shell *Archivolva alexbrownii* Red Sea, 16 mm. © Sven Kahlbrock.

Golden Egg Shell *Crenavolva aureola* WP, 11 mm. Feeds on *Acanthogorgia* gorgonians. Yellow to orange or reddish shell, mantle with white dots, foot and siphon are black.

Chiapponi's Egg Shell *Crenavolva chiapponii* WP: Philippines, Brunei, 10 mm. Synonymised with *C. aureola* based on DNA studies of specimens from Indonesia. Kept as a separate species for now. © Brian R. Mayes.

Groves' Egg Shell *Crenavolva grovesi* WP: Japan to Philippines and New Caledonia, 12 mm. Feeds on sea whips. Translucent white or yellowish mantle, domed papillae. Left: © Chay Hoon, right: © Tsu Soo Tan.

Guido's Egg Shell *Crenavolva guidoi* WP: Philippines to Brunei and Australia, 10 mm. Feeds on sea whips. Orange-brown to mauve shell, large white papillae. © Brian R. Mayes.

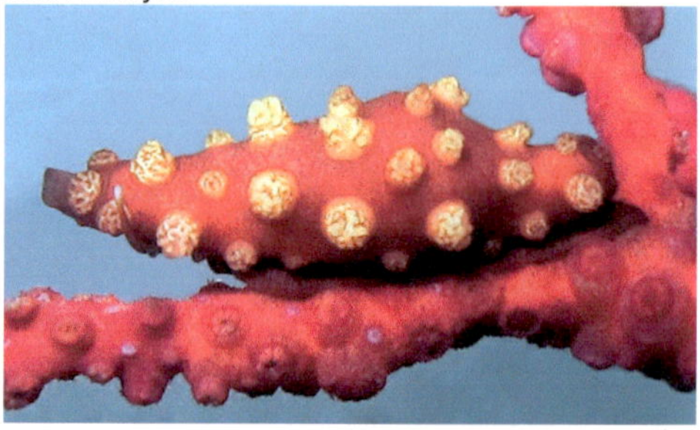

Matsumiya's Egg Shell *Crenavolva matsumiyai* Japan to Papua New Guinea and Solomons, 14 mm. Feeds on *Echinogorgia*. Red or brown mantle, large papillae. Right: © Sue Churchill & Rogan Draper, left: © J. Nishio.

Striated Egg Shell *Crenavolva striatula* IWP, 16 mm. Feeds on *Echinogorgia* sea fans. Red and yellow color forms, both with 3 pale transverse bands on the shell. © Philippe & Guido Poppe www.poppe-images.com

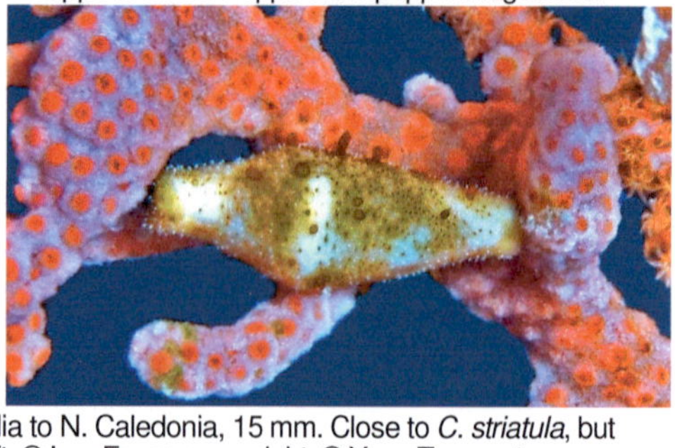

Marbled Egg Shell *Crenavolva marmorata* WP: Australia to N. Caledonia, 15 mm. Close to *C. striatula*, but possesses no distinct transverse bands on the shell. Left: © Luc Faucompre, right: © Yves Trenevet.

Leopard Egg Shell *Crenavolva leopardus* WP: Japan to Indonesia, Brunei, 12 mm. Pale to red-orange shell without distinct marking, with fine striation over the dorsum. Mantle with small dark spots. © Brian R. Mayes.

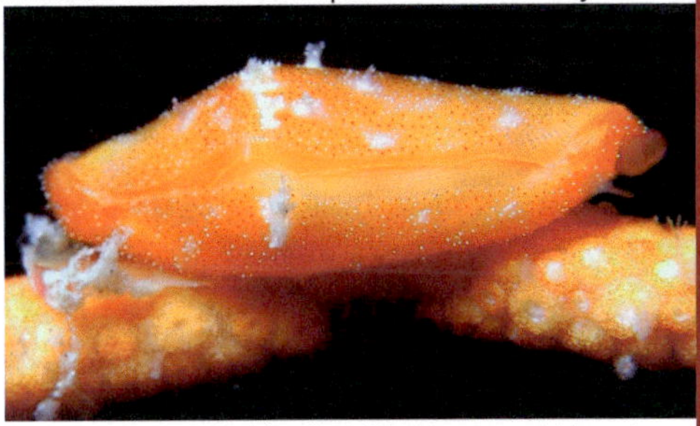

Trail's Egg Shell *Crenavolva traillii* WP, 16 mm. Red or yellow mantle, small white dots. Transverse middorsal pale band is rare. Feeds on several species of gorgonians. © Philippe & Guido Poppe www.poppe-images.com.

Tinted Egg Shell *Crenavolva tinctura* WP: Japan to Australia, 13 mm. Feeds on several species of gorgonians, including *Paraplexaura*, *Subergorgia*, *Echinomuricea*. Shell with angular profile, orange at the tips.

Tinted Egg Shell *Crenavolva tinctura* (continued) Mantle with red-brown spots and tiny white dots, often united into transverse islands. Siphon and foot transparent to reddish.

Odd Egg Shell *Crenavolva* sp. WP: Indonesia, Brunei, 12 mm. © Arto Jarvenpaa.

Glassy Egg Shell *Crenavolva vitrea* WP: Japan, Philippines, 14 mm. © Yui Kawahara.

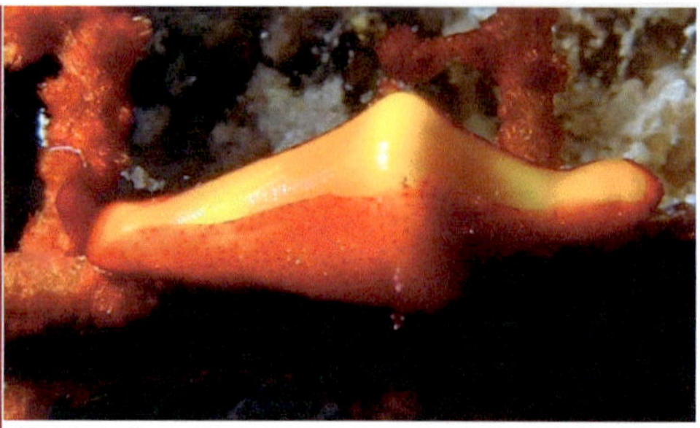

Martin's Egg Shell *Crenavolva martini* IO: E. Africa, Red Sea (1st report), 14 mm. Photo: © Sven Kahlbrock.

Tosa's Dissona *Dissona tosaensis* WP: Japan, 15 mm. Mantle with red spots and white dots. Photo: © J. Nishio.

Allyn Smyth' Egg Shell *Cuspivolva allynsmithi* WP: Japan, Korea, Brunei, 11 mm. Similar to *C. cuspis* (below) but differs by the less colorful rose-grey shell with tips of terminals creamy yellow. © Brian R. Mayes.

Pointed Egg Shell *Cuspivolva cuspis* WP: Japan to Vietnam, Brunei, 12 mm. Hosts on *Euplexaura* sp., *Ellisella* sp. Recognizable inflated shell with pointed posterior tip, mantle with dark and white spots. © Brian R. Mayes.

Draper's Egg Shell *Cuspivolva draperi* IP: East Africa to Japan and Australia, 18 mm. Shell with angular profile and wavy pattern, visible through the mantle. Mantle with yellow-brown spots. © Stephen D.A. Smith.

Howland's Egg Shell *Cuspivolva howlandae* Philippines to Australia, 10 mm. © Stephen D.A. Smith.

Ostheimer's Egg Shell *Cuspivolva ostheimerae* WP: Australia to Japan, 8 mm. © Brian R. Mayes.

Platypus Egg Shell *Cuspivolva platysia* WP: Japan to Australia, 12 mm. Shell with orange areas on the tips. Mantle and foot with black spots, black siphon. © Stephen D.A. Smith (left), © Brian R. Mayes (right).

Beautiful Egg Shell *Cuspivolva pulcherrima* WP: Philippines, Brunei, 16 mm. Shell with a cloudy violet and white pattern. Hosted on *Euplexaura* sp. gorgonian sea fan. © Brian R. Mayes.

Elongated Egg Shell *Cuspivolva* sp. Indonesia, 12 mm. ID tentative. © Charles E. Rawlings. ID tentative.

Tiger Egg Shell *Cuspivolva tigris* WP: S. Korea to Indonesia, New Caledonia and Vanuatu, 17 mm.

Tiger Egg Shell *Cuspivolva tigris* (cont'd) Yellow shell, mantle with transverse black bands. Yellow foot with rounded black spots. Feeds on *Euplexaura* sea fans. Probably imitates toxic nudibranch *Phyllidia ocellata*.

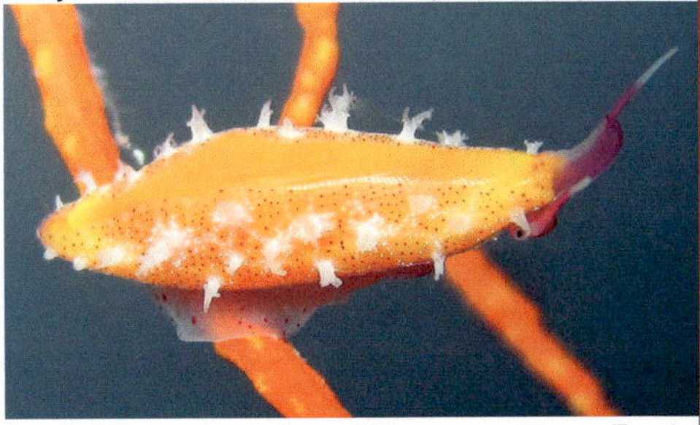

Queensland Egg Shell *Cuspivolva queenslandica* IWP: South Africa to Australia and Singapore, 12 mm. Feeds on *Pseudopterogorgia* sea fans. Translucent mantle with small maroon dots. Photos © Sue Churchill & Rogan.

Azuma's Egg Shell *Dentiovula azumai* IWP: South Africa to Japan, 10 mm. Yellow to pale maroon to match host's color, mantle with dark red spots and prominent papillae. © Johji Nishio.

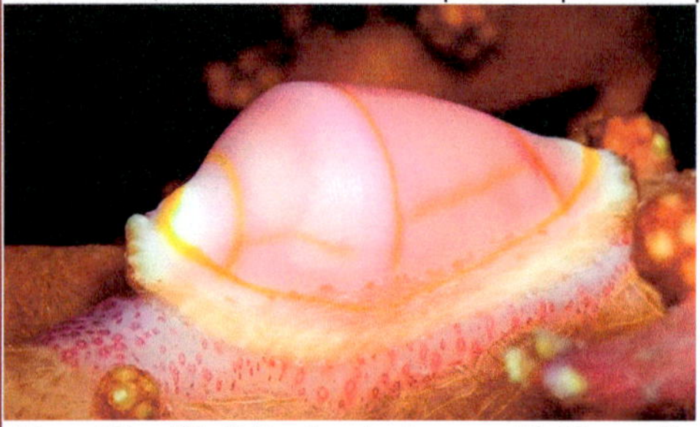

Dorsal Egg Shell *Dentiovula dorsuosa* IP, 18 mm. Feeds on *Siphonogorgia* and *Chironephthya octocorals*. The shell is white in color with yellow lines. Foot with pink rings. © Brian R. Mayes.

Dorsal Egg Shell *Dentiovula dorsuosa* (cont'd) Pink or yellowish mantle with small red dashes and white tapering papillae. Right: © Nick Hobgood, left: © Rokus Groeneveld.

Eizo's Egg Shell *Dentiovula eizoi* IP: East Africa to Central Pacific, 12 mm. Feeds on *Acanthogorgia* sea fan. Mantle with brown dots, prominent branching papillae. Left: © I. Bogachenko, right: © Dave Johnson.

Eizo's Egg Shell *Dentiovula eizoi* (continued) Several color morphs from yellow to maroon are matching the colors of host gorgonians. ©JJ.

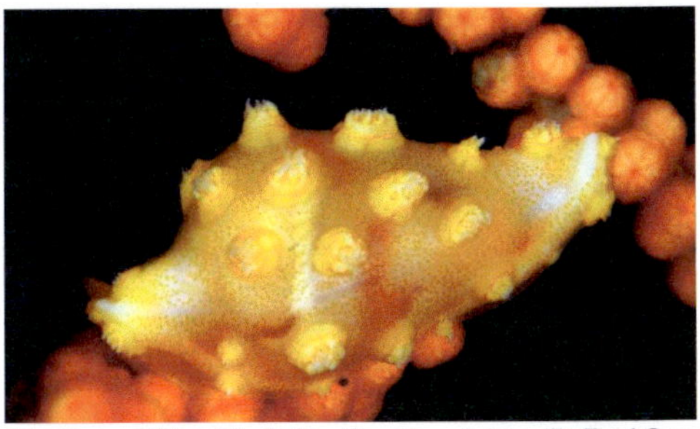

Shortened Egg Shell *Dentiovula colobica* IP: Red Sea to Japan, 11 mm. Shell with humped profile, translucent yellow mantle, prominent papillae are matching the host's polyps. © Johji Nishio.

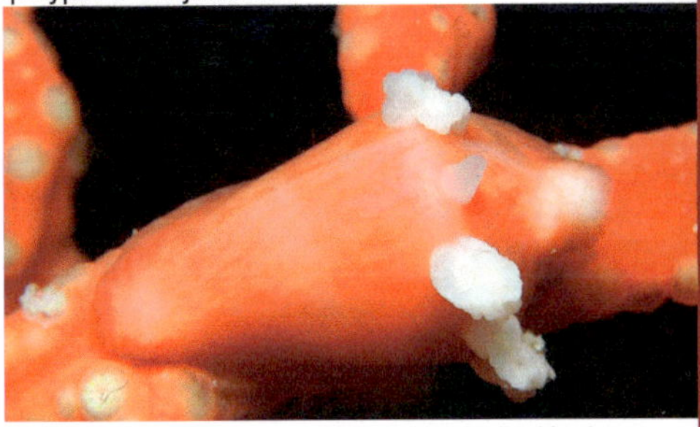

Hora's Egg Shell *Dentiovula horai* WP: Brunei to Japan and Philippines, 18 mm. Rhomboid shell with sharp angulation at the shoulder, cream to mauve or reddish. © Philippe & Guido Poppe www.poppe-images.com.

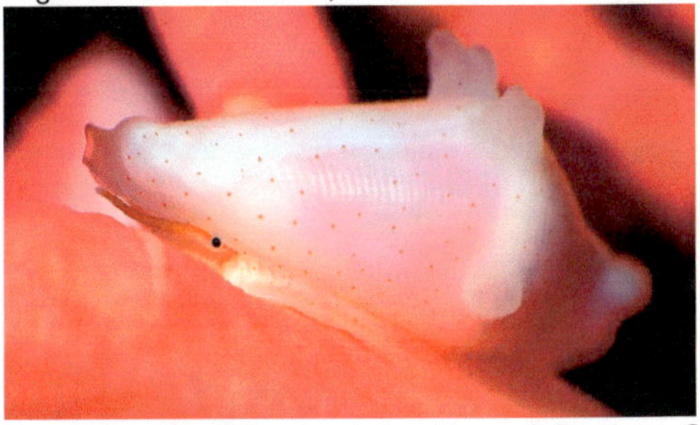
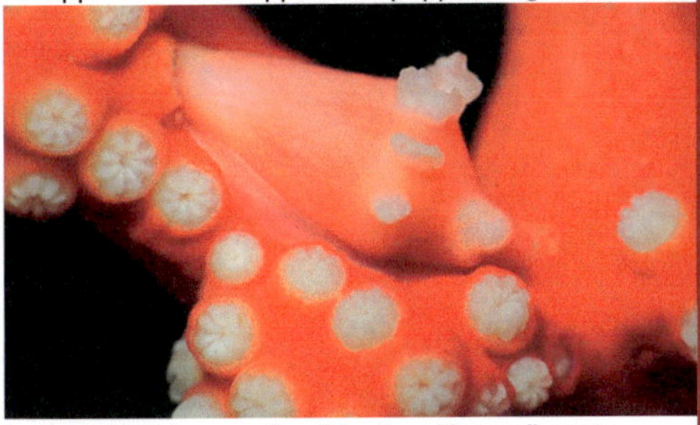

Hora's Egg Shell *Dentiovula horai* (cont-d) Feeds on *Solenocaulon* false sea fan. Mantle with small spots, white, pink or reddish in color. Often long papillae across the shoulder. © Brian R. Mayes.

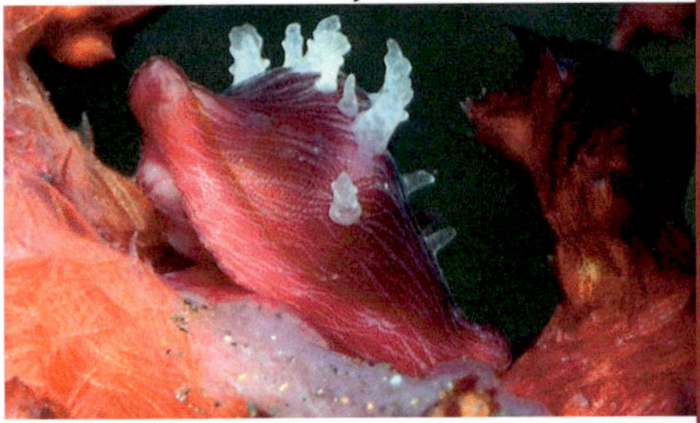

Silky Egg Shell *Dentiovula mariae* IP: Madagascar to Papua New Guinea, 8 mm. The shell is pale purplish to mauve in color. Feeds on *Chironephthya* soft corals. Left: © Steven Ng, right: ©JJ.

Silky Egg Shell *Dentiovula mariae* (cont'd) Similar to *D. dorsuosa* (above), but much smaller, with angular, less rounded profile. © Brian R. Mayes.

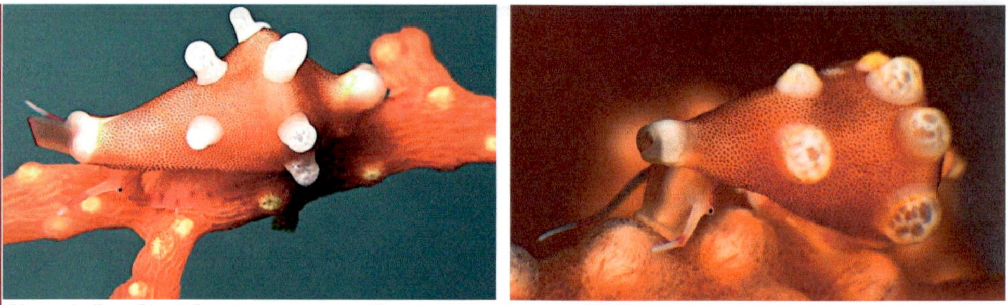

Masao's Egg Shell *Dentiovula masaoi* WP: Brunei to Japan and Vanuatu, 18 mm. The shell with sharp angulation at the shoulder, cream to orange or reddish. © Sue Churchill & Rogan Draper (left), © Steven Ng (right).

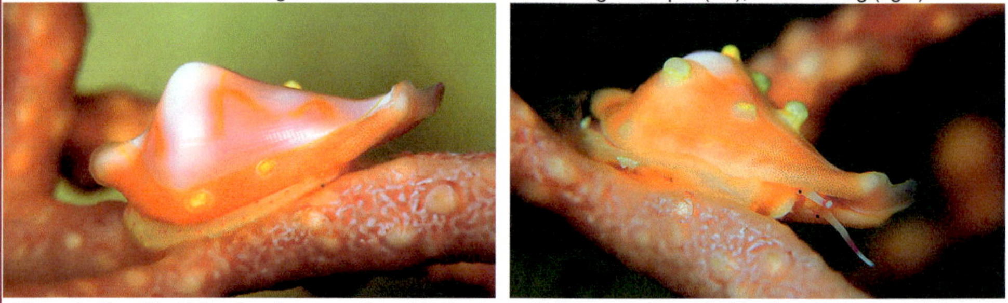

Masao's Egg Shell *Dentiovula masaoi* (cont'd) Feeds on *Siphonogorgia* corals. Mantle with tiny dark dots and large papillae, imitating coral polyps in form and color © Brian R. Mayes.

Constellate Egg Shell *Primovula astra* WP: Japan, Brunei, New Guinea, 10 mm. Pinkish white striated shell with bright orange spots. Translucent mantle with white papillae and orange dots. © Brian R. Mayes.

Violet Egg Shell *Primovula formosa* IWP, 16 mm. Feeds on *Euplexaura* sea fans. Easily identified by bright purple color. Left: © Rokus Groeneveld, right: © Tsu Soo Tan.

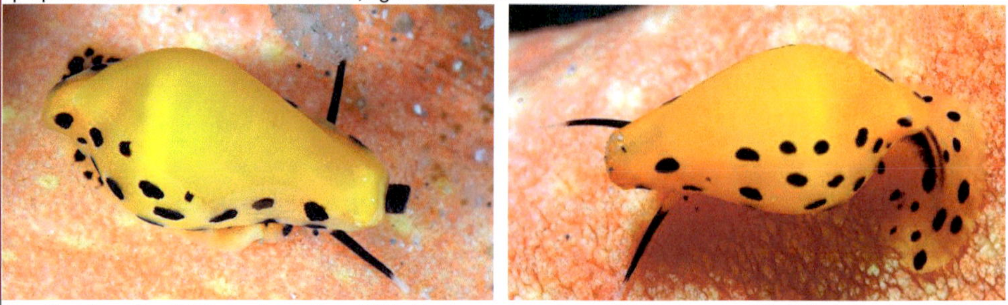

Panther Egg Shell *Primovula panthera* IWP: Madagascar to Japan, Philippines and N. Caledonia, 8 mm. Feeds on *Paraminabea* corals. The shell is yellow in color, translucent mantle with dark spots. © Johji Nishio.

Boomerang Egg Shell *Primovula roseomaculata* WP, 10 mm. Feeds on *Scleronephthya* soft corals. Shell with 3 rows of orange boomerangs, visible through the mantle. Left: © Dennis Teck Lee Liew, right: © Johji Nishio.

Rosewater's Egg Shell *Primovula rosewateri* IWP: South Africa to West Pacific, 15 mm. The shell with more or less angular profile, smooth or calloused, with orange terminals. Feeds on a variety of octocorals. ©JJ (right).

Rosewater's Egg Shell *Primovula rosewateri* (cont'd) White mantle with large rounded spots and scattered yellow conical papillae, the pattern and color depend on host. © Brian R. Mayes.

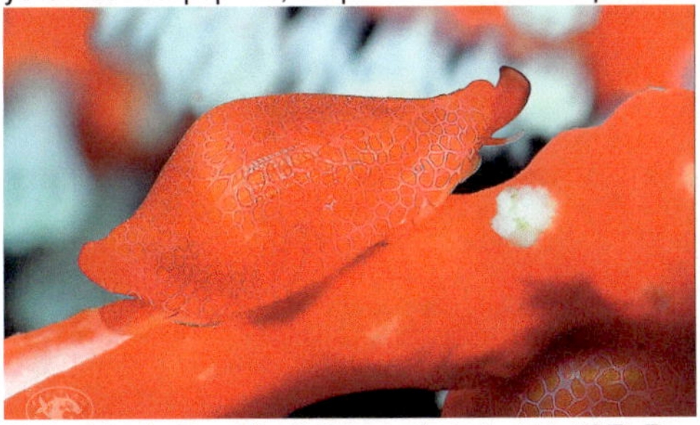
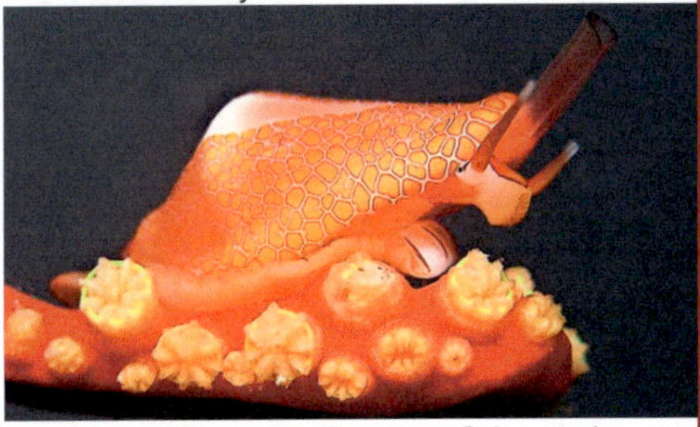

Tadashige's Egg Shell *Primovula tadashigei* WP: Brunei to Japan, PNG, 14 mm. Feeds on *Solenocaulon* false sea fan. Right photo © Sue Churchill & Rogan Draper, left: © Brian R. Mayes.

Seven-Spots Egg Shell *Rotaovula septemmacula* Red Sea and E. Africa to Japan, 10 mm. Feeds on different gorgonians. Shell with 7 brown spots. Right (Japan): © Johji Nishio, left (Red Sea): © Sven Kahlbrock.

Hirohito's Egg Shell *Rotaovula hirohitoi* IWP: Japan, Reunion, Philippines, 10 mm. © Yoichi Sato.

Robertson's Egg Shell *Quasisimnia robertsoni* WCP: Japan to Marquesas, 33 mm. © E.G de Suduiraut.

Korkos' Egg Shell *Prosimnia korkosi* Red Sea, 13 mm. Spindle-shaped shell with pointed terminals. Greenish to light-brown mantle with darker dots and tapering papillae. White foot with dark dots. © S. Kahlbrock.

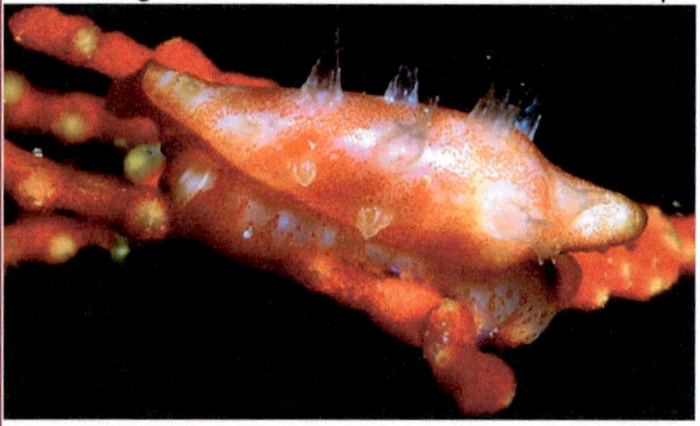

Hepca's Egg Shell *Prosimnia hepcae* Red Sea, 6 mm. Red form. Found on *Melithaea sinaica* sea fan (red morph). © Sven Kahlbrock. Papillae mimic the host sea fan polyps.

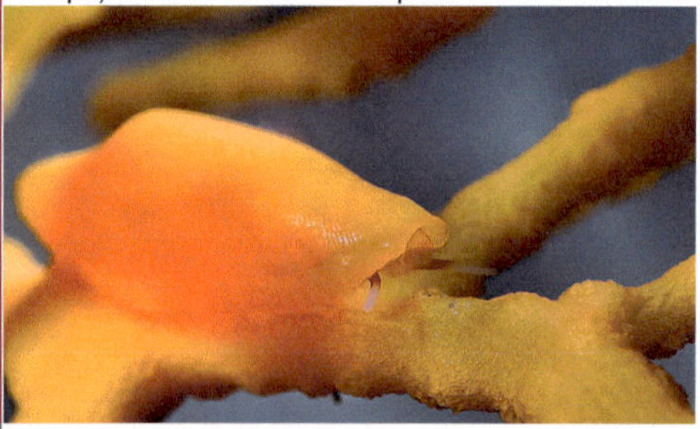

Hepca's Egg Shell *Prosimnia hepcae* (continued) Red Sea, 6 mm. Feeds on yellow *Melithaea sinaica* sea fan. Yellow shell, translucent mantle, white papillae. Left: © Rafi Amar, right: © Sven Kahlbrock.

a-h. **Dragon Egg Shell** *Prosimnia draconis* WCP, 16 mm. Shell with a strong transverse ridge.
j-i. **Semper's Egg Shell** *Prosimnia semperi* IWP, 15 mm. Shell without a strong transverse ridge.
Species delimitation in this group needs to be reconsidered with molecular data. Draper & Churchill (2014) treated them as variations of *P. semperi*.
Photos: a, b, h - ©JJ, c - © Bjarte Hekonsen, f,g - © Yves Trenevet,
j,i - © Sven Kahlbrock.

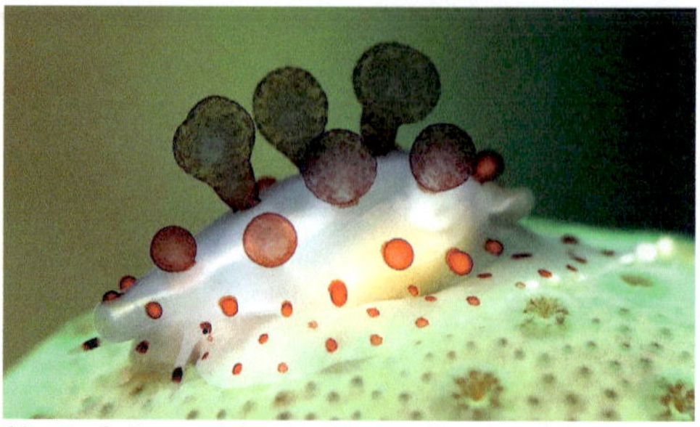

Umbilical Egg Shell *Calpurnus verrucosus* IWP: East Africa to Solomon Islands, 40 mm. Feeds on leather corals *Sarcophyton* sp. Right: juvenile, © Anke Wentzler.

Pink-Mouth Egg Shell *Ovula costellata* IWP: E. Africa to WP, 50 mm. Feeds on *Sarcophyton* leather corals.

Isibasi's Egg Shell *Ovula isibasii* IWP: E. Africa to WP, 40 mm. Deep water (>90 m). © E.G de Suduiraut.

Great Egg Shell *Ovula ovum* IP: Red Sea and East Africa to Japan and French Polynesia, 120 mm. Feeds on leather corals *Lobophytum* and *Sinularia*. White shell, black mantle with white or yellow papillae.

Great Egg Shell *Ovula ovum* (continued) Juvenile with yellow-tipped papillae.

Great Egg Shell *Ovula ovum* (continued) Juveniles are imitating the toxic *Phyllidia* nudibranchs. Right: ©Dave Johnson.

Pacific Pedicularia *Pedicularia pacifica* IP, 8 mm. Feeds on *Stylaster* lace corals. Coloration variable from red to purple and yellow and matches the host color. ©JJ.

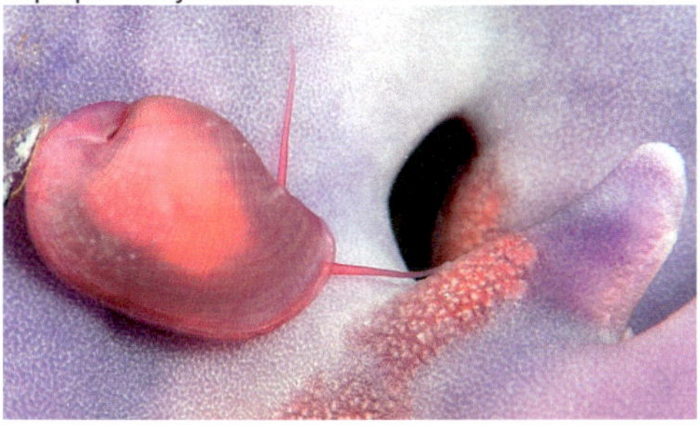

Vanderland's Pedicularia *Pedicularia vanderlandi*, WCP: Indonesia to Marshall Islands, 8 mm. Feeds on *Distichopora* lace corals. Species delimitation in this group needs to be reconsidered with molecular data. ©JJ.

Netted Milky Egg Shell *Procalpurnus lacteus* IP: Red Sea to Solomon Islands, 24 mm. Feeds on leather corals *Sinularia* sp. White shell, mantle with brown network pattern. Grey siphon with white spots. ©JJ

Spotted Milky Egg Shell *Procalpurnus semistriatus* IP: Red Sea to Papua New Guinea, 21 mm. Feeds on leather corals. Similar to *P. lacteus*, differs by the siphon and foot patterns. Photo on the left: ©SJ.

Spotted Milky Egg Shell *Procalpurnus semistriatus* (cont'd) Cephalic pattern, photo: Stephen D. A. Smith.

Adamson's Egg Shell *Pseudocypraea adamsonii* IP Reunion to Marshalls, 11 mm. ©SJ.

Tiny Purple Trivia Shell *Purpurcapsula exigua* IP: Red Sea, Japan to Hawaii, 5 mm. Shell with transverse ridges. Mantle with a few yellowish papillae. left: (Guam) - © Bob Abela, right: (Red Sea) © Rafi Amar.

Starry Trivia Shell *Trivia* sp. WCP to Marshall Islands, 10 mm. Shell with transverse ridges. Tan to red mantle with conical papillae, each on star-like base. Like all *Triviidae*, feeds on compound tunicates. ID Tentative. ©JJ.

Starry Trivia Shell *Trivia* sp. (cont'd) Siphon red-brown, translucent downward part. Foot pinkish with pattern of white lines. Reddish tentacles with white dots. ©JJ.

Slug-Like Trivia Shell *Trivia* sp. CP: Marshall Islands, 11 mm. Black mantle with white compound papillae, decorated with yellow spots. Mimic *Colpodaspis thompsoni* opistobranch. ©JJ.

Branching Trivia Shell *Trivia* sp. CP: Marshall Islands, 10 mm. White shell with transverse ridges. White mantle with dark margins, very long tree-like papillae. ©JJ.

Orange-Spotted Trivia Shell *Trivia* sp. WCP, 10 mm. Mantle with orange spots. ©JJ. ID tentative.

White-Netted Trivia Shell *Trivia* sp. WP: Indonesia, 10 mm. Translucent foot with white lines. ID tentative.

Brown-Mantle Trivia Shell *Trivia* sp. WP: Philippines, 10 mm. Orange-brown mantle. ID tentative.

Pale Trivia Shell *Trivia* sp. CP: Marshalls, 10 mm. White mantle, pale orange spots. ID tentative. ©JJ.

Rough Trivia Shell *Trivirostra scabriuscula* IP: Red Sea, Mozambique to Guam, 13 mm. White shell with transverse ridges. Red-brown mantle, siphon and tentacles are covered by white dots.

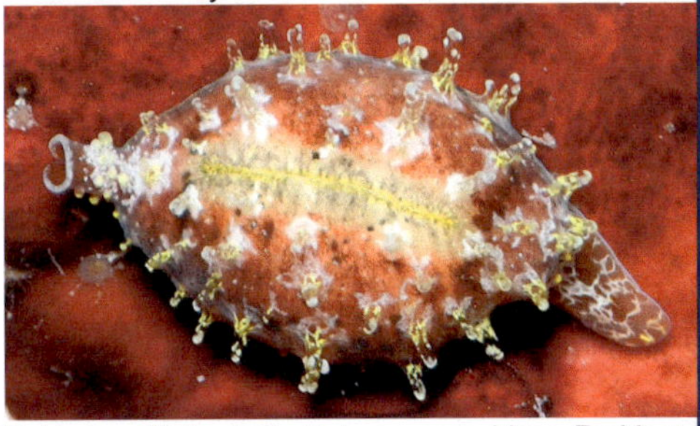

Sea Nymph Island Trivia Shell *Trivirostra spioinsula* IWP, 11 mm. White shell with transverse ridges. Red foot with opaque white lines. Left: © Bob Abela, right - ©JJ.

Barley Trivia Shell *Trivirostra hordacea* IP: Reunion to Hawaii, 10 mm. Dark grey mantle. © Y. Trenevet.

White-Covered Trivia Shell *Trivirostra* sp. CP: Micronesia, 9 mm. ©JJ.

TRIVIIDAE

Orange Trivia Shell *Trivirostra* sp. CP: Marshall Islands, 10 mm. ID tentative, possibly a form of *Trivirostra spioinsula* with fewer tentacles on the mantle. ©JJ.

ERATOIDAE

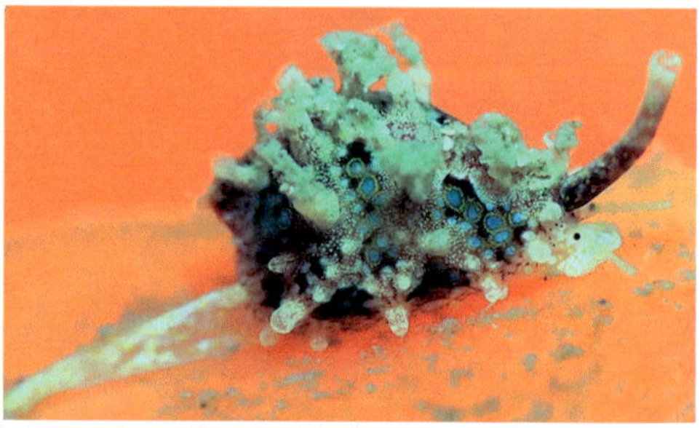

Yellow-Netted Trivia Shell *Proterato lachryma* WP, SP: Japan, Australia, New Zealand, 8 mm. Smooth white or cream shell with brown bands posteriously. Mantle with blue ocelli and knobby papillae. © Brian R. Mayes.

Red-Spotted Eratoena *Eratoena sulcifera* IP: East Africa to Central Pacific, 5 mm. Greenish finely granulated shell. Mottled mantle with white papillae and areas around them. ©JJ.

Yellow-Spotted Erato *Erato* sp. WP, 10 mm. ©JJ. Grey to brown mantle covered by large nipple-like papillae, decorated by yellow and orange spots.

Sandwich Islands Eratoena *Eratoena sandwichensis* Hawaii, 4 mm. Bluish shell, white-dotted mantle. ©SJ.

Tiny Eratoena *Eratoena nana* IP: E. Africa to Guam, 5 mm. Mantle with dark dots. © Bob Abela.

Velvet Snail *Coriocella hibyae* IO: Maldives, Andaman Isl, 10 cm. Thin shell is hidden under the fleshy dark blue or dark brown mantle. Mimics the sponge with its scattered round dark spots. Feeds on ascidians.

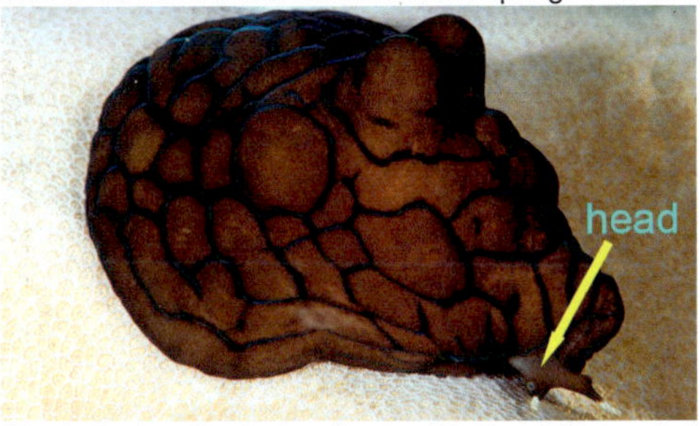

Black Coriocella *Coriocella nigra* IP: East Africa to Polynesia and Hawaii, 120 mm (shell 40 mm). Small thin shell is completely covered by the mantle. Dark brown head with white-tipped tentacles. Feeds on ascidians.

Black Coriocella *Coriocella nigra* (cont'd) Coloration variable, from yellow-brown to dark blue. Photo on the right (©JJ) - a small red-brown juvenile with numerous dark dots.

Black Coriocella *Coriocella nigra* (cont'd) Yellow-brown morph. Reported from Western and Central Pacific. Coriocella species are similar in appearance to sponges or dead corals. ©JJ.

Violet Coriocella *Coriocella* sp. WP: Philippines, 3 cm. ID tentative, probably juvenile form of *Coriocella nigra*. Mimic sponges, imitating sponge osculum with apically black appendages.

Lamellaria species are predators that prefer to feed on colonial ascidians. These cryptic animals are very well camouflaged, mimicking the sponges or ascidians. Like other members of Velutinidae family they bear a thin internal shell.

a-i: Sponge mimic *Lamellaria* sp. come in various colors that may be just a few species.

Photos h-i - ©JJ, g - ©SJ.

Species-level identification of all *Lamellaria* species is difficult from the photographs. Cryptic animals, they are rarely seen by divers or underwater photographers.
a-f: Sponge mimic *Lamellaria*
g-i: Ascidians mimic *Lamellaria*
Photos: a-h - ©JJ.

a-h: Ascidians mimic *Lamellaria* of various colors and forms. All Lamellaria species are feeding on ascidians, including the sponge imitating ones. Species-level identification from the photographs is impossible.
j: Marseniopsis sp., Indonesia. j: Marseniopsis sp., Philippines.
Photos: a-h - ©JJ.
Identification tentative.

Lavender-Mouth Conch *Canarium erythrinum* IP: Red Sea, East Africa to Marshall Islands, 50 mm. Yellow to red shell with striated lips. Like other conchs they use narrow muscular feet for moving in jerks or jumps. ©JJ.

Plicate Conch *Canarium labiatum* IWP: East Africa to West Pacific, 50 mm. Cream to white shell, banded with pale brown. Living animal greenish with opaque white mottling.

Maculated Conch *Canarium maculatum* CP: Marshall Islands to Hawaii, Polynesia, 40 mm. Shell of a dingy white color, with obscure brown spots. Living animal white with green speckles. Good jumper with 1 m+ jumps. ©JJ.

Micro Conch *Canarium microurceus* WCP, 32 mm. Smooth cream to white shell with brown maculations. Living animal green with opaque white spots.

Changeable Conch *Canarium mutabile* IP: Red Sea, East Africa to Marshall Islands, 45 mm. Shells are mostly whitish in color, but can be decorated with white and brown bands or longitudinal stripes. ©JJ.

Wilson's Conch *Canarium wilsonorum* IP: East Africa to Guam and Hawaii, 31 mm. Small conch, shell variable in color, often with thin transverse bands. ©JJ.

Red-Mouth Conch *Canarium scalariforme* IP, 25 mm. White striated shell with longitudinal ribs. ©JJ.

Mauritian Conch *Conomurex decorus* IP: E. Africa to Singapore, 80 mm. Smooth shell with pink aperture.

Blood Mouth Conch *Conomurex luhuanus* WP, 80 mm. Smooth cone-shaped shell with faint brown banding and mostly orange aperture (hence the common name). Photo on the left: ©JJ.

Minute Conch *Dolomena minima* WP to Fiji, 45 mm. Shell with a wing-shaped rounded outer lip. Smooth shell, dingy brownish or brown with white banding. Animal grey-green with white speckles.

Pigeon Conch *Dolomena plicata* IP: Red Sea to Maldives, 77 mm. Dull white spirally striated shell.

Swainson's Conch *Dolomena swainsoni* WP, Vietnam, 80 mm. Shell fusiformly ovate, lip winged.

Variable Conch *Dolomena variabilis* WCP: Indonesia and Philippines to Marshall Islands, 60 mm. Medium-size conch, shell white to dingy brownish with transverse mottled bands.

Variable Conch *Dolomena variabilis* (continued) Claw-like serrated along one edge operculum, typical for conchs (left). Eyes protruding, the one along the anterior canal, the other under the notch of outer lip (right). ©JJ

Riband-Marked Conch *Doxander vittatus* WCP: South Chine Sea to Fiji. © Ria Tan.

Plough Conch *Euprotomus aratrum* WP, 115 mm. Orange shell opening, flared lip. © Ria Tan.

Bubble Conch *Euprotomus bulla* WP to Samoa, 90 mm. Thick elliptical shell, flared outer lip with a prong-like posterior expansion. Cream-brown with fawn mottling outside. Aperture orange with white margins.

Bubble Conch *Euprotomus bulla* (continued) Juveniles glossy red-brown to yellow, often without mottling. Eyes are bordered with distinct blue pigment.

Imperial Conch *Euprotomus aurisdianae* IWP: East Africa to Philippines, rare find at Marshall Islands, 95 mm. Outside of shell grey-brown with fawn mottling, ventral side heavily glazed.

Vomer Conch *Euprotomus vomer* WP: Australia to Japan, 100 mm. Shell with expanded outer lip, fawn to redbrown with white speckles. Animal red-orange with white spots. ©SJ

Hawaiian Conch *Euprotomus hawaiensis* Hawaii, 110 mm. Live animal red with white spots. ©SJ

Humped Conch *Gibberulus gibberulus gibbosus* WCP, absent in Hawaii, 7 cm.

Humped Conch *Gibberulus gibberulus gibbosus* (cont'd) Roughly fusiform shell with asymmetrically coiled inflated spire. Aperture white or purple. Live animal white with brown speckling.

Chiragra Spider Conch *Harpago chiragra* East Africa to Polynesia and Marshall Islands, 330 mm. Shell with six recurved elongate spines, extending from the aperture. Photo on the left: ©SJ, right: ©JJ.

Harpago chiragra (cont'd) As the fingers develop, they remain hollow for a while until they fill in. ©SJ.

Swan Conch *Labiostrombus epidromis* WP to New Caledonia, 95 mm. Outside of the shell pale-brown.

Dog Conch *Laevistrombus canarium* IWP: Sri Lanka to Melanesia, Australia and Japan, 105 mm. Pale brown to yellow-brown shell with a a flared, thick outer lip.

Turtle Conch *Laevistrombus turturella* WP: 115 mm. Thick heavy shell with flared lip. © Ria Ran.

Vanikoro Conch *Laevistrombus vanikorensis* SP, 55 mm. © Uwe Weinreich.

Smooth Spider Conch *Lambis lambis* IP: East Africa to Marshall Islands, 275 mm. Shell with seven more or less straight fingers (marginal spikes) of varying length. Photo on the left: juvenile, fingers undeveloped.

Smooth Spider Conch *Lambis lambis* (continued) ©SJ. Adult, ventral view.

Millipede Spider Conch *Lambis millepeda* IWP, 200 mm. Shell with ten short curved fingers.

Orange Spider Conch *Lambis crocata* IWP, 205 mm. Anterior canal expanded in a very long and slender spike. Close to *L. truncata* (below), but with curved siphonal canal spine. ©JJ.

Scorpion Spider Conch *Lambis scorpius* IWP, 220 mm. Shell with alternating black and white transverse lines on on each side of the aperture. Aperture with seven long fingers. Photo on the right: ©JJ.

Truncate Spider Conch *Lambis truncata* IP: East Africa to Marshall Islands, 420 mm. The largest of spider shells. Close to *L. lambis* (above) but with a more squarish outline. Outer lip with six hollow fingers. ©SJ.

Freckled Conch *Lentigo lentiginosus* IP: East Africa to Marshall Islands, 105 mm. Heavy thick shell, flaring outer lip posteriorly expanded. Outer lip with pale brown blotches. Aperture margins with yellow and black dots.

Butterfly Conch *Lentigo pipus* East Africa to Guam and Marshall Islands, 85 mm. Transversely striated shell. Proboscis with red and white bands. Left: ©JJ, right: © Bob Abela.

Robust Conch *Margistrombus robustus* WP: Japan to Indonesia, 58 mm. Cone-shaped brown-banded shell.

Seventh Conch *Margistrombus septimus* WP, 70 mm. Outer lip extending on the spire, white-banded shell.

Finger Spider Conch *Ophioglossolambis digitata* IP: East Africa to Indonesia and Samoa, 180 mm. Shell yellow-brown, expanded cheek divided into ten rounded segments.

Widest Mouth Pacific Conch *Sinustrombus latissimus* WP: Philippines to Melanesia, 226 mm. Recognizable shell with rounded expansion of the outer lip much higher than the spire.

Laciniate Conch *Sinustrombus sinuatus* IWP: Thailand to Philippines, Marshall Islands and Guam, 170 mm. Expansion of the outer lip nearly as high as the spire. Photo on the left: juvenile.

Bull Conch *Sinustrombus taurus* IP: Seychelles to Marshall Islands, 130 mm. Shell with two pointed extensions of the outer lip, one of them very long. Photo on the left: ©SJ, right: ©JJ.

Fragile Conch *Terestrombus fragilis* WCP, 53 mm. Shell of variable coloration from red and yellow to dingy brown with faint banding. Locally common inhabitant of shallow and deep lagoons.

Fragile Conch *Terestrombus fragilis* (continued) Living animal green with white speckles and rounded spots. Eyestalks also green with white rounded spots.

Little Auger Conch *Terestrombus terebellatus* IWP: Red Sea to Fiji and Solomon Isl., 54 mm. ©SJ.

Thersite Conch *Thersistrombus thersites* WP, 170 mm. © Uwe Weinreich.

Three-Knobbed Conch *Tricornis tricornis* Red Sea to Somalia, 170 mm. Outer lip w/acute-angled expansion.

Toothed Conch *Tridentarius dentatus* IP: East Africa to Hawaii, 65 mm.

Toothed Conch *Tridentarius dentatus* (continued) Outer lip with 3 or 4 teeth on anterior margin (photo on the right, marked with arrow). Photo on the left: juvenile, teeth on the lip are just forming.

Terebellum Conch *Terebellum terebellum* IP: East Africa to Marshall Islands and Polynesia, 75 mm. Thin outer lip without an anterior notch. Shell coloration variable. Proboscis with red and white pattern. ©JJ.

Black-Mouth Moon Snail *Mammilla melanostoma* IP: E. Africa to Polynesia, 51 mm. Globular white shell with 3-4 pale brown spiral bands. Feeds on bivalves, drilling a hole in the prey's shell by means of enzymes. ©JJ.

Simi Moon Snail *Mammilla simiae* IP: Red Sea, East Africa to Polynesia, 40 mm. White shell with 3 spiral brown bands. Animal white-yellowish. ©JJ.

Breast-Shaped Moon Snail *Mammilla mammata* WP, 35 mm. Shell with faint brown bands. © Ria Tan.

Spider Moon Snail *Natica arachnoidea* IP: E. Africa to Marshall Isl., 26 mm. Rare orange shell color. ©JJ.

Spider Moon Snail *Natica arachnoidea* (cont'd) Thick shell with low spire, variable in color. Often with brown maculations that run diagonally from the spire towards base. ©JJ.

Solid Moon Snail *Natica fasciata* IWP: E. Africa to Fiji, 30 mm. Mottled brown animal, dingy-orange shell.

Kawamura's Moon Snail *Natica kawamurai* WP: Japan, Philippines, 20 mm. ©JJ.

Striated Moon Snail *Natica pluvialis* WP: Philippines, 22 mm. Cream shell with pattern of axial orange-brown lines. Grey-brownish mantle skirt with white bands and red dots.

Starry Moon Snail *Natica stellata* IWP: Indonesia, Papua New Guinea to Japan, 45 mm. Yellow-brown or orange shell with a darker central band and 2 spiral rows of white blotches. White black-blotched mantle skirt.

Starry Moon Snail *Natica* cf. *stellata* WP: Philippines, 25 mm. Pale brown shell with 2 spiral rows of white blotches. Animal with white bands and red-pigmented head shield. ©JJ.

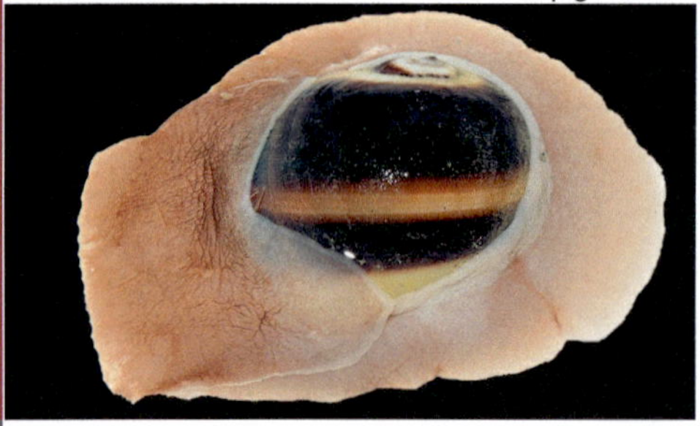

Yolk Moon Snail *Natica vitellus* IWP: Madagascar to Japan and New Caledonia, 50 mm. Yellow-brown to creamy shell with 1 to 3 broad and distinct spiral brown bands. Right: © Ria Tan.

Deceptive Moon Snail *Natica pseustes* IWP: South Africa to Japan, Philippines and Marshall Islands, 15 mm. White shell with pattern of rusty-brown wavy lines. White animal with red dots. ID tentative. ©JJ.

Deceptive Moon Snail *Natica pseustes* (cont'd) Moon snail hides under sand when disturbed. ©JJ.

Elegant Moon Snail *Naticarius concinnus* WP: Japan, Marshalls, 16 mm. ID Tentative. © Jay Lord.

Butterfly Moon Snail *Naticarius alapapilionis* IP: East Africa to Fiji, 35 mm. Light brown or tan shell with 5 spiral red and white lines. Animal dingy tan or translucent with white lines.

Butterfly Moon Snail *Naticarius* cf. *alapapilionis* WP: Indonesia, Philippines, 40 mm. Yellow-brown shell with 5 spiral brown and white lines. Animal pink or red with white bands. ©JJ.

Butterfly Moon Snail *Naticarius* cf. *alapapilionis* (cont-d) Voracious predators, moon snails are rasping the soft parts of molluscan prey after drilling a circular hole in the shell of the prey.

Oriental Moon Snail *Naticarius orientalis* IWP: East Africa to Japan, Philippines and New Guinea, 37 mm. Cream to brown-orange shell.

Oriental Moon Snail *Naticarius orientalis* (cont'd) ©JJ. Animal is bright red with radial white stripes. Hides under sand during the day, hunts on bivalves and gastropods at nighttime.

China Moon Snail *Naticarius onca* IP: East Africa to Japan and Fiji, 35 mm. Yellow-brown to cream shell with rows of dark brown spots. Translucent mantle skirt with opaque white radial bands, head shield with red pigment.

China Moon Snail *Naticarius onca* (continued) Pale form. ©JJ.

Waisted Moon Snail *Naticarius zonalis* IWP, 28 mm. Dull yellow shell, red head shield. © Ria Tan.

Hepatic Moon Snail *Neverita didyma* IWP, 90 mm. Orange to brown shell, umbilical callus (highlighted by arrow) dark brown, with a strong groove. © Ria Tan.

Gualtieri's Moon Snail *Notocochlis gualtieriana* IWP, 30 mm. Photo on the left: ©JJ. Creamy shell with a pattern of axial brown lines. White animal with brown dots.

Red-Dotted Moon Snail *Notocochlis* sp. CP: Marshall Islands, 18 mm. White shell with 2 yellow spiral bands and axial wavy lines. White animal with red dots.

Handsome Moon Snail *Notocochlis venustula* IWP: Reunion; Philippines, 13 mm. Yellowish shell with 3 spiral bands of thick brown patterned bands. White animal with brown dots.

Nebulose Moon Snail *Notocochlis cernica* IWP: East Africa to Japan, South Pacific, 22 mm. Creamy to tan shell with a pattern of wavy brown lines. Animal with long antennae, totally covered by red or brown dots.

Nebulose Moon Snail *Notocochlis* cf. *cernica* WP, 15 mm. Animal with blurry pattern. ID tentative.

Orange-Dotted Moon Snail *Notocochlis* sp. WP, 20 mm. Brown shell, orange-dotted animal. ID tentative.

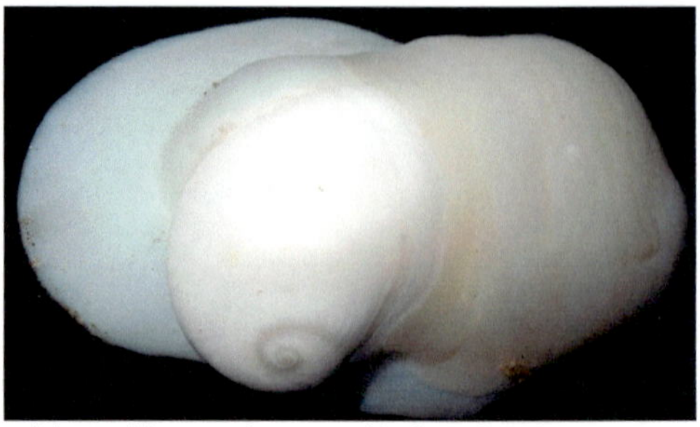

Egg-White Moon Snail *Polinices albumen* IWP: Reunion, Mozambique to Japan and Indonesia, 68 mm. Compressed white shell, with a very flat spire, white mantle skirt. © Ria Tan.

Golden Moon Snail *Polinices aurantius* WP, 58 mm. White to chestnut shell. © David Massemin.

Golden Moon Snail *Polinices aurantius* (cont'd) Shell with relatively high spire. ©JJ.

Cuming's Snail *Polinices cumingianus* WP, 55 mm. © Philippe & Guido Poppe www.poppe-images.com.

Tiger Moon Snail *Paratectonatica tigrina* 40 mm. IWP: Red Sea, Madagascar to Australia and Japan. Pearshaped shell covered with red-brown dots. © Ria Tan.

White Moon Snail *Polinices mammilla* IP: Red Sea, East Africa to Marshall Islands, 68 mm. Pear-shaped thick shell is totally white, as well as the animal. ©JJ.

Areolate Moon Snail *Tanea areolata* IWP: Red Sea, East Africa to Philippines and Japan, 18 mm. Yellow-orange shell with pattern of a wavy pale lines.

Lined Moon Snail *Tanea lineata* IWP: Mozambique, S. Africa to Japan, 52 mm. White shell with pattern of axial orange-brown lines. White or pale-brown mantle skirt. © Ria Tan.

Undulate Moon Snail *Tanea undulata* IWP: Red Sea, East Africa to Philippines and Japan, 35 mm. This color morph with 3 pointing forward spiral bands was recently treated as a separate species, *T. euzona*. Photo on the left: ©JJ.

Undulate Moon Snail *Tanea undulata* WP, 35 mm. White or creamy shell with pattern of a wavy, axial orangebrown thick lines. Head shield with 3 distinct black bars.

Violet Moon Snail *Tectonatica violacea* WCP: Philippines to Australia and Marshall Islands, 25 mm. Pear-shaped shell covered with red-orange spots, arranged in rows. White mantle skirt with dark red spotting. ©JJ.

FIG SHELLS - FICIDAE

Common Fig Shell *Ficus ficus* IWP, 165 mm. Pearshaped shell with a low spire. Nocturnal animal.

Threaded Fig Shell *Ficus filosa* WP: Australia to Japan, 120 mm. Depth>150 m. © E.G de Suduiraut.

Variable Fig Shell *Ficus variegata* IP: S. Africa to Japan and Singapore, 120 mm. Mottled brownish-red and white mantle. Additional genetic data needed to confirm species delimitation with *Ficus ficus*. © Ria Tan.

Underlined Fig Shell *Ficus subintermedia* IWP: Mozambique to Philippines and New Guinea, 130 mm. Fawn in color, with irregular brown marks. Left (juvenile) - © Marty Kiser. Right - © E.G de Suduiraut.

LITTORINIDAE

Mangrove Periwinkle *Littoraria scabra* IP, 40 mm. Lives on intertidal rocks, feeding on alga and bacteria.

Undulate Periwinkle *Littoraria undulata* IP: East Africa to Hawaii, 25 mm. Found on intertidal rocks.

RISSOINIDAE

Ambiguous Rissoina *Rissoina ambigua* IP, 8 mm. White shell with red-brown pigment or bands. ©JJ.

Ribbed Rissona *Rissoina* sp. Melanesia, 10 mm. White ribbed shell, opaque white animal. ©JJ

Three-Toothed Zebina *Zebina tridentata* IP: Reunion to Marshalls, 12 mm. Outer lip usually with three denticles on the lower half. ID tentative. ©JJ.

Apple Tun *Malea pomum* IWP: Red Sea, East Africa to Polynesia and Hawaii, 90 mm. Thick polished shell with low spire, pale brown with darker areas, banded with ribs. White animal with violet-brown insertions.

Apple Tun *Malea pomum* (cont'd) Like other family members, hunts at night on *Holothuria* species.

Ribbed Tun *Tonna allium* IWP: East Africa to Japan, Melanesia and New Guinea, 131 mm.

Ribbed Tun *Tonna allium* (cont-d) Cream or fawn shell, darker on the ribs, spiral ribs without squarish brown blotches. Photo on the right: feeding on *Holothuria*. White animal with violet-brown lines and mottling.

Lischke's Tun *Tonna lischkeana* IP: East Africa to Hawaii, 215 mm. Cream shell, spiral ribs with squarish brown blotches. White animal with violet-brown lines and mottling. Photo on the right: juvenile.

Channeled Tun *Tonna canaliculata* IWP: East Africa to Japan, Melanesia and Australia, 145 mm. Cream or fawn shell, ribs with narrow interstices, low spire.

Channeled Tun *Tonna canaliculata* (cont-d) Foot brown with opaque white areas and dark margin, much larger than its shell.

Black-Mouthed Tun *Tonna melanostoma* WCP, 307 mm. Feeds on large Holothuria species. ©SJ.

Partridge Tun *Tonna perdix* IWP: Red Sea, East Africa to Japan and Polynesia, 227 mm.

Partridge Tun *Tonna perdix* (cont'd) Large shell with rounded spiral ribs. Purplish brown outside, pale in grooves, white markings on the ribs. ©JJ.

Partridge Tun *Tonna perdix* (cont'd) Brown animal with white mottling. Left: © David Burdick, right: © Sebastien Vasquez - feeding on paralysed *Actinopyga echinites* sea cucumber.

Thersites Frog Shell *Bufonaria thersites* WP, 60 mm. © David Massemin. MNHN Koumac 2018.

Roughest Frog Shell *Bursa asperrima* IP: East Africa to Philippines, 60 mm. Shell with plain white aperture.

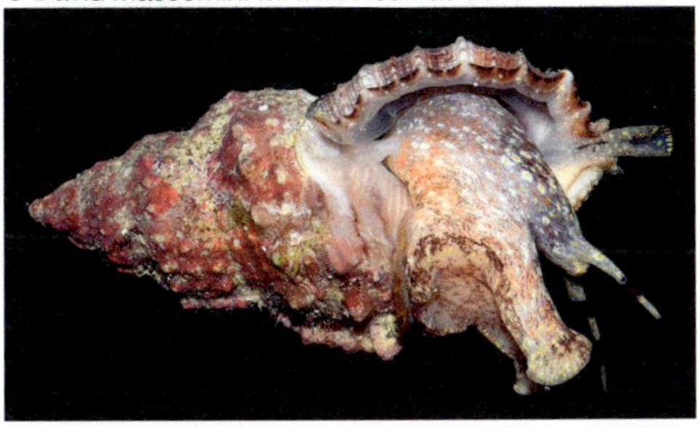

Similar Frog Shell *Bursa affinis* WCP: Okinawa, Philippines, Guam and French Polynesia, 80 mm. (Based on Sanders et al. 2017). Ovate shell with 2 varices, animal red-purplish, tentacles with black and white bands.

Granular Frog Shell *Bursa granularis* IWP: Mozambique to the Red Sea, Vietnam, Japan, the Philippines, 60 mm. Shell relatively tall, narrow. Very common. *B. granularis* and *B. affinis* are difficult to distinguish morphologically.

Barty Frog Shell *Bursa bufonia* IWP: E. Africa to Polynesia, Japan and Hawaii, 10 cm. Thick shell with coarse knobbed sculpture. Grey animal with round white spots, some with orange dot in the centre. ©JJ.

Tall Frog Shell *Bursa condita* IWP: Maldives to Polynesia, 110 mm. Very tall, narrow shell. ©JJ.

Blood-Stained Frog Shell *Bursa cruentata* IWP, 50 mm. Shell with dark-red spots on the inner lip. ©JJ.

Lamarck's Frog Shell *Bursa lamarckii* IWP, 80 mm. Shell w/long posterior siphonal canals, dark aperture.

Triton Frog Shell *Bursa ranelloides* IP, 85 mm. Shell similar to tritons. Animal green with white spots.

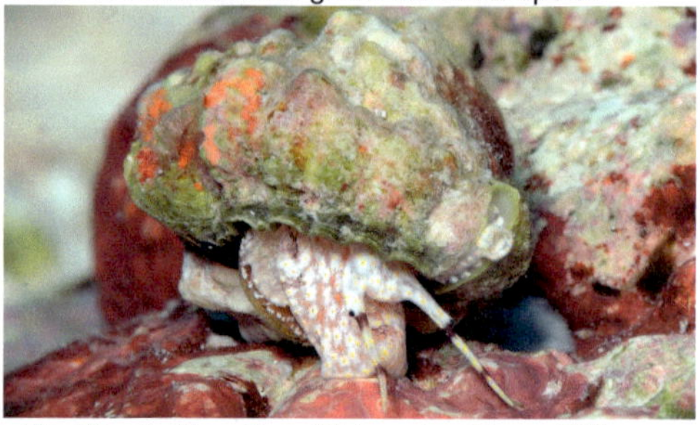

Wine-Mouth Frog Shell *Bursa rhodostoma* IP: E. Africa, Red Sea to Hawaii and Polynesia, 35 mm. Differs from other Bursa in having very short posterior siphonal canals. Often with dark red aperture.

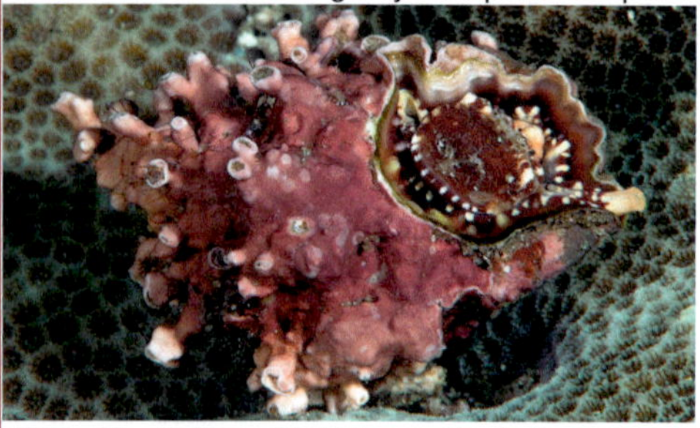

Wine-Mouth Frog Shell *Bursa rhodostoma* (cont'd) Pale mauve animal with white spots.

Yellow Mouthed Frog Shell *Bursa tuberosissima* WP, 62 cm. Recognised by yellow aperture. ©JJ

Rosy Frog Shell *Bursa rosa* IWP, 55 mm. Shell with pinkish mauve aperture, two rows of long posterior siphonal canals. Animal pinkish mauve with white spots. Photos: left: ©JJ right: ©SJ

Black Mouthed Frog Shell *Bursa venustula* Marshalls to Polynesia, 41 m. Shell with both lips black. ©JJ

Fijian Frog Shell *Bursina fijiensis* WCP to Polynesia, 65 mm. White to fawn shell, animal w/pale orange spots.

Small Frog Shell *Bursina ignobilis* IWP: E. Africa to N. Caledonia, 65 mm. Animal mottled orange and tan.

Bohol Frog Shell *Tutufa boholica* WP: Philippines, 120 mm. Shell with pinkish mauve aperture and animal.

Giant Frog Shell *Tutufa bubo* IP: East Africa to the Philippines and Polynesia, 337 mm. Shell with 1 row of denticles inside the outer lip. White animal with brown mottling. Photo on the left: juvenile, ©JJ, right: adult, ©SJ.

Red-mouthed Frog Shell *Tutufa rubeta* IWP: Rйunion to Taiwan, Australia and Polynesia, 250 mm. Aperture reddish in color, with 2 rows of denticles inside the outer lip. Animal dark mauve in color. Right photo: ©SJ.

Common Bonnet *Casmaria erinaceus* IP: Red Sea, E. Africa to Melanesia and Micronesia, 79 mm. Reddish brown, outer lip thickened, with wide dark-brown bars and 3-5 denticles, animal white to grey-brownish.

Common Bonnet *Casmaria erinaceus* (cont'd) Like other family members, night hunter. Feed on echinoderms (sea urchins). First squirt neurotoxic saliva to paralyse the prey. ©JJ (left)

Bob Lehman's Bonnet *Casmaria boblehmani* WCP, 55 mm. Nodulose shell with spiral bands.

Unicolor Bonnet *Casmaria unicolor* IO: R. Sea, Oman 72 mm. Cream shell with inflated upper whorl profile.

Heavy Bonnet *Casmaria ponderosa* IP: Red Sea, East Africa to Hawaii, 100 mm. Smooth or nodulose shell (2 distinct forms). Outer lip with 7-9 spines. White or grey-purplish animal. Photo on the left: ©JJ.

Heavy Bonnet *Casmaria ponderosa* (cont'd) Nocturnal, active predator. Preys on sea urchins, paralyzing them with neurotoxic saliva. Left: © Petr Vorotnikov, Red Sea, Dahab.

Zigzag Bonnet *Casmaria turgida* WP: Japan and Philippines to Vanuatu, 100 mm. Cream to yellow shell with characteristic zigzag color pattern. Animal white or creamy.

Giant Helmet *Cassis cornuta* IP: E. Africa to Hawaii, 410 mm. Shell with 3 or 4 spiral rows of large tubercles, shoulder tubercles much longer than the others. Preys on the Crown-of-Thorns starfish *Acanthaster planci*. ©JJ.

Giant Helmet *Cassis cornuta* (cont'd) Ventral side cream, yellow or orange ©SJ.

Red Helmet *Cypraecassis rufa* IP, 200 mm. Reddish shell, very thick outer lip. © Philippe Bourjon.

Banded Bonnet *Phalium bandatum* WP: Japan and Philippines to Australia, 140 mm. Shell with upturned siphonal canal, outer lip with 2-3 spines, 5 interrupted brownish spiral bands. Photos: left ©SJ right (eggs!) ©JJ.

Banded Bonnet *Phalium bandatum* (cont'd) Animal light-brown, preys on sea urchins. Nocturnal.

Grey Bonnet *Phalium glaucum* IP: E. Africa and Madagascar to Japan and Melanesia, 147 mm.

Grey Bonnet *Phalium glaucum* (cont-d) Globose grey shell with a low spire, outer lip with 3-4 sharp spines. Animal red-brown, nocturnal. Preys on sea urchins.

Japanese Bonnet *Semicassis bisulcata* IWP: Red Sea, East Africa to Japan, Philippines and Australia, 85 mm. Globular white, cream or fawn shell, usually with 4 to 5 spiral bands of brown spots. Preys on sea urchins.

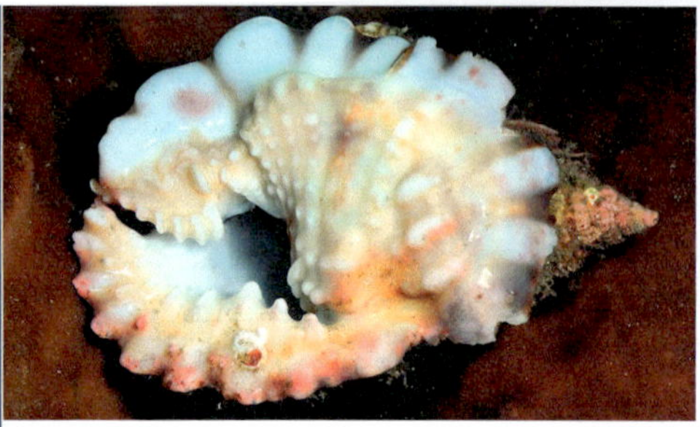

Common Distorsio *Distorsio anus* IP: Red Sea, East Africa to Hawaii and Polynesia, 100 mm. Shell with distorted aperture, sinuous inner lip. Short and strongly recurved siphonal canal.

Common Distorsio *Distorsio anus* (cont'd) interesting note: "anus" in Latin could mean "ring" or "old woman." So the name really might not mean what some of you think.

Common Distorsio *Distorsio anus* (cont'd) Red-orange animal with rounded white spots. Nocturnal, hides during the day under rocks and rubble. Photos: ©SJ, ©JJ (left, right).

Reticulate Distorsio *Distorsio reticularis* IWP: Persian Gulf to Taiwan, Philippines and Australia, 94 mm. Shell with narrow and distorted aperture, long and slightly recurved siphonal canal right ©JJ.

Reticulate Distorsio *Distorsio reticularis* (cont'd) Red-orange animal with rounded white spots, preys on the polychaete worms (*Phyllochaetopterus*) ©JJ.

Triton's Trumpet *Charonia tritonis* IP: Red Sea, East Africa to Japan, Hawaii and Polynesia, 490 mm. Shell with a tall spire, glossy, with purple brown markings.

Triton's Trumpet *Charonia tritonis* (cont'd) Preys on large echinoderms, including ill famous coral-eating Crown of Thorns starfish, *Acanthaster planci*. This juvenile is eating a young cushion star, *Culcita novaeguineae*. ©SJ.

Triton's Trumpet *Charonia tritonis* (cont'd) Frequently over-collected species. Collections and sale of *Charonia tritonis* now banned in some countries (Australia, India).

Lesser Girdled Triton *Gelagna succincta* IWP: Red Sea, East Africa to Central Pacific, 80 mm. Usually hard to recognize due to long paddles of soft periostracum ("coat") that cover the striped shell. Animal red-orange. ©JJ.

Tiny Triton *Distorsomina pusilla* IP: Comoro Islands to Japan, Australia and Marshalls, 12 mm. ©JJ.

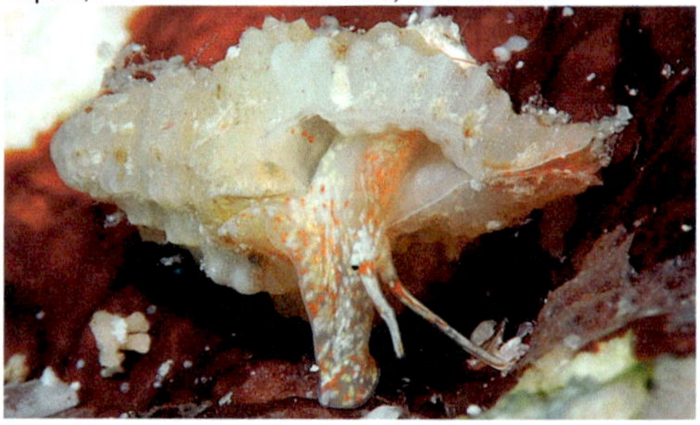

Tiny Triton *Distorsomina pusilla* (cont-d) Smallest tonnoidean. Very narrow shell with unusually weakly distorted coiling. Animal with red-yellow-white mottling. Left (depositing eggs, aquarium) - © Bob Abela. Right: ©JJ.

Shortneck Triton *Gutturnium muricinum* IWP: E. Africa to Polynesia, Hawaii, also Atlantic, 80 mm. Ovate shell with thick glossy white inner and outer lips. Preys on bivalves. ©JJ.

Tadpole Triton *Gyrineum gyrinum* IP: Somalia, India to Japan and French Polynesia, 50 mm. Recognizable yellow shell with dark-brown banding. White animal with red and yellow spotting. ©JJ.

Purple Gyre Triton *Gyrineum lacunatum* IP: Red Sea, East Africa to Japan and Hawaii, 20 mm. Shell variably pink, brown and white, with a mauve aperture. Yellowish animal with a mauve spotting. ©JJ.

Rosy Gyre Triton *Gyrineum roseum* WP: Japan to Polynesia, 30 mm. Yellow, or red w/yellow nodules. ©JJ.

Black-Spotted Triton *Lotoria lotoria* IP: E. Africa to Japan, Australia and Polynesia, 160 mm. ©JJ.

Black-Spotted Triton *Lotoria lotori*a (cont'd) Large orange-brown shell with 2 large black spots on the inner lip. Red-orange animal with dark red spots. Photo on the right - juvenile. ©JJ.

Aquatile Hairy Triton *Monoplex aquatilis* Circumtropical: E. Africa and Red Sea to Hawaii and Costa Rica, 120 mm. Pale orange to brown shell, pale orange aperture. Pink animal with red rounded spots. Preys on bivalves. ©JJ.

Dwarf Triton *Monoplex comptus* IP: East Africa to Philippines and Marshall Islands, 40 mm. Shells vary in color from grey to beige, orange and even to black. Animal with red to dark-brown spots.

Jeweled Triton *Monoplex gemmatus* WP: Japan to Australia and Polynesia, 35 mm. Small narrow shell, narrow spiral cords, yellow-brown to medium brown. Animal white, with red rounded spots. ©JJ

Pure Triton *Monoplex mundus* IP: E. Africa to Japan and Australia and Solomons, 40 mm. White or pale-brown shell with a white aperture. Pink animal with red spots. ©JJ.

Nicobar Hairy Triton *Monoplex nicobaricus* IP: East Africa to Japan and Hawaii, 103 mm. Grey shell, orange aperture, lips with white rim. Animal grey with red-brown round spots all over. Preys on bivalves and other shells.

Common Hairy Triton *Monoplex pilearis* IP: Red Sea, East Africa to Japan, Hawaii and Australia, 140 mm. Elongated tan or pale brown shell with bright red aperture. White animal with mauve round spots. Preys on bivalves.

Orange-Mouthed Triton *Ranularia gutturnia* IWP East Africa to Marshalls, 100 mm. Shell with low spire, thick glossy orange lips and very long siphonal canal. Animal white to yellowish with round pink spots. right - ©JJ.

Slender Triton *Ranularia exilis* IWP, 60 mm. Red Sea, E. Africa to Philippines and N. Caledonia, 60 mm.

Necklace Triton *Ranularia monilifera* IWP, 60 mm. Moderately long, recurved anterior siphonal canal.

Forgotten Triton *Ranularia oblita* WP: Philippines to Australia and N. Guinea, 65 mm. © Marty Kiser.

Three-Lined Triton *Ranularia trilineata* IO: Red Sea, E. Africa, 90 mm. Strongly sculptured brown&white shell.

Pear Triton *Ranularia pyrum* IP: Red Sea, 130 mm. Orange strongly sculptured pear-shaped shell with orange lips and recurved anterior siphonal canal. Tan animal with red-brown spots. Left: © Bob Abela, right: ©SJ.

Bibbey' Triton *Septa bibbeyi* WCP: Philippines to Guam, 50 mm. © Bob Abela.

Red Sea Triton *Septa marerubrum* IO: Red Sea, 50 mm. Red shell with yellow transverse ridges on lips.

Red Triton *Septa rubecula* IP: East Africa to Hawaii, 50 mm. White animal with large red spots. ©JJ.

Black-Striped Triton *Septa hepatica* IP: Reunion to Japan and Tahiti, 65 mm. ©JJ.

Wide-Lipped Triton *Turritriton labiosus* IWP, 50 mm. Short and wide shell, recurved siphonal canal. ©JJ.

Echinaster's Apicalia *Apicalia echinasteri* WP: Australia to Japan and Solomons, 9 mm. Ovate, solid, transparent grey or white shell. White animal with long retractible proboscis. Ectoparasite on *Echinaster* starfish. ©JJ.

Spotted Annulobalcis *Annulobalcis maculatus* WP, 7 mm. Elongated transparent conical-shaped glossy shell. Red-brown body with large yellow spots. Ectoparasite on crinoids. Photo on the left ©JJ.

EULIMIDAE

Capillaster Goodingia *Goodingia capillastericola* IWP: Red Sea (Aqaba Bay), Indonesia, 7 mm. Short transparent glossy shell. Animal with white, brown-striped body. Ectoparasite on crinoids. Photos: ©SJ,

Purple Annulobalcis *Annulobalcis yamamotoi* WP, 8 mm. Found on crinoids. © Stephen D. A. Smith.

Wine Annulobalcis *Annulobalcis vinarius* WP, 8 mm. Parasite on crinoids. Animal red with white dots. ©SJ.

Fire Urchin Snail *Echineulima asthenosomae* WP: Australia, Indonesia, Philippines, 12 mm. Semitransparent conical-shaped shell. Parasitic on *Asthenosoma* sea urchins. Photo on the left: with eggs.

Robust Echineulima *Echineulima robusta* WP, 15 mm. Ectoparasite on *Diadema* sea urchins. ©SJ.

Red-Dotted Hypermastus *Hypermastus* sp. CP, 10 mm. Ectoparasite on burrowing sea urchins. ©JJ.

White Crinoid Snail *Eulima* sp. WP: Indonesia, Philippines. Elongated transparent conical-shaped glossy shell. Ectoparasite on crinoids. ID tentative. Right photo: ©SJ.

Double-Band Eulima *Eulitoma* sp. WP: Philippines, Indonesia, 14 mm. White conical-shaped shell with double spiral line of orange color. Ectoparasite on burrowing sea urchins.

Sandfish Snail *Melanella* sp. WP: Philippines, 10 mm. Ectoparasite on *Holothuria* sp. cucumber. ID tentative.

Cuming's Melanella *Melanella cumingii* IP, 45 mm. Ectoparasite on sea cucumbers. ©JJ.

Green Melanella *Melanella dufresnii* IP: Reunion to Marshalls and Hawaii, 25 mm. Elongated transparent conical-shaped glossy shell. Animal green, proboscis with yellow outline. ID tentative. ©JJ

Red-Dotted Melanella *Melanella* sp. CP, Melanesia, 24 mm. Transparent shell. White animal with red dotting, tentacles with red stripes. Ectoparasite on sea cucumbers, including *T. anax*, *S. chloronotus*. ©JJ

Yellow-Bordered Melanella *Melanella* sp. CP, Melanesia, 25 mm. Transparent shell. Animal with a double pair of yellow lines running up each tentacle and a yellow anterior foot margin. Ectoparasite on sea cucumbers. ©JJ.

Spotted Melanella *Melanella* sp. CP, Melanesia, 15 mm. Host unknown. © Christina Sylvester.

Milky Melanella *Melanella* sp. CP, Melanesia, 18 mm. Host unknown. © Christina Sylvester.

Yellow-Mustache Melanella *Melanella* sp. CP, 12 mm. On sea cucumbers. © C. Sylvester.

Green-Yellow Melanella *Melanella* sp. CP, Melanesia, 10 mm. © Christina Sylvester.

Long-Mouth Melanella *Melanella teinostoma* WCP, 36 mm. Ectoparasite on sea cucumbers. ©JJ.

Brown Niso *Niso brunnea* WP: Japan to Indonesia, 18 mm. Dark-brown glossy shell, host unknown. ©JJ.

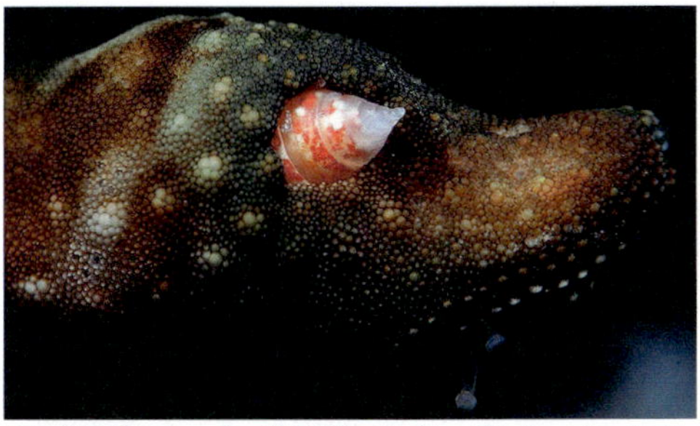

Starfish Boring Snail *Stilifer linckiae* IP: Reunion to Hawaii and Marshalls, 7 mm. Bores into the arms of sea stars, leaving only the tip of its spire exposed. © Christina Sylvester.

Yellowish Starfish Snail *Parvioris fulvescens* IWP: Thailand - N. Caledonia, 6 mm. © D. Massemin.

Brown Parvioris *Parvioris* sp. CP, Melanesia, 20 mm. Brown body with yellow streaks. © C. Sylvester.

Green Parvioris *Parvioris* sp. CP, Melanesia, 20 mm. Green body with white streaks. © C. Sylvester. IT.

Green-Spotted Parvioris *Parvioris* sp. CP, Melanesia, 20 mm. Orange tentacles. © C. Sylvester. IT.

Opaque Parvioris *Parvioris* sp. CP, Melanesia, 17 mm. Parasite on starfishes. ©SJ. IT.

Red-Striped Peasistilifer *Peasistilifer nitidula* IP, 5 mm. Ectoparasite on sea cucumbers. ©JJ

Sea Urchin Snail *Vitreobalcis* sp. CP, Melanesia, 7 mm. Ectoparasite on sea urchins. © C. Sylvester.

Starfish Cap Snail *Thyca ectoconcha* IO, 8 mm. Parasites on *Linskia multifora* starfish. © P. Bourjon.

EULIMIDAE

Crystalline Starfish Snail *Thyca crystallina* IP, 6 mm. Parasitic on *Linckia* starfish, attaching to the host starfish by the snout and inserting the proboscis into the radial canal (Waren, 1983). Right photo: ©SJ.

Green-Spotted Ophiura Snail *Stilapex montrouzieri* WP, 9 mm. An ectoparasite of *Macrophiotrix* cf. *longipeda* brittle star. Translucent shell, animal with green spots. NMNH expedition photos. Right: © Arthur Anker.

Brittle Star Zebra Snail *Stilapex zebra* Parasitic on the oral side of *Ophiothrix* brittle star. © S. Kahlbrock.

VANIKORIDAE

Mantis Shrimp Snail *Caledoniella montrouzieri* IWP: Persian Gulf and Madagascar to American Samoa and Japan, 8 mm. An obligate ectoparasite of gonodactylid stomatopods.

Cancellate Vanikoro *Vanikoro cancellata* IP: East Africa to Marshalls, 25 mm. ©JJ

Ribbed Vanikoro *Vanikoro gueriniana* WP, 15 mm. White shell with thin dentate ribs.

Sea Urchin Cap Shell *Hipponix prionocidaricola* WCP: Japan to Marshalls, 15 mm. Smaller males (blue arrows) and females (yellow arrows) on the spines of *Phyllacanthus* sea urchin. Photo (left, right): ©SJ, ©JJ.

False Cup-and-Saucer *Cheilea equestris* Circum-tropical, 40 mm. Firmly attached to the substrate. ©JJ.

Conical Hoof Shell *Sabia conica* IWP, 45 mm. Commonly found attached to other seashells. ©JJ.

Moderate Tube Shell *Caecum* cf. *modestum* WP: Indonesia, Philippines to Vanuatu, 1,7 mm. Smooth translucent subcylindrical tube, slightly curved. White animal, round operculum. Intertidal, under small rocks.

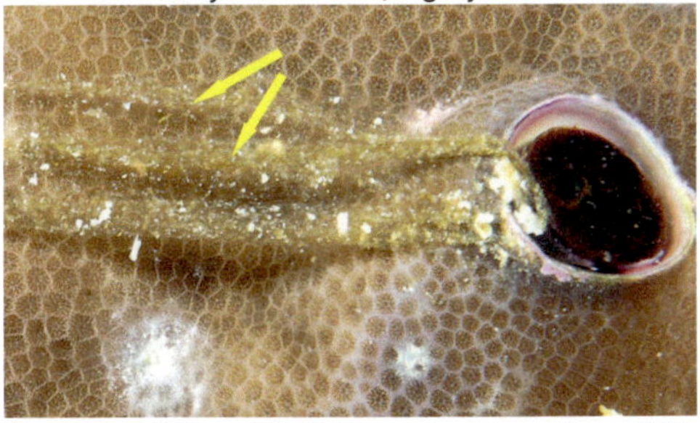

Great Coral Worm Shell *Ceraesignum maximum* WP, 20 mm (diameter). Mucus feeding net highlighted.

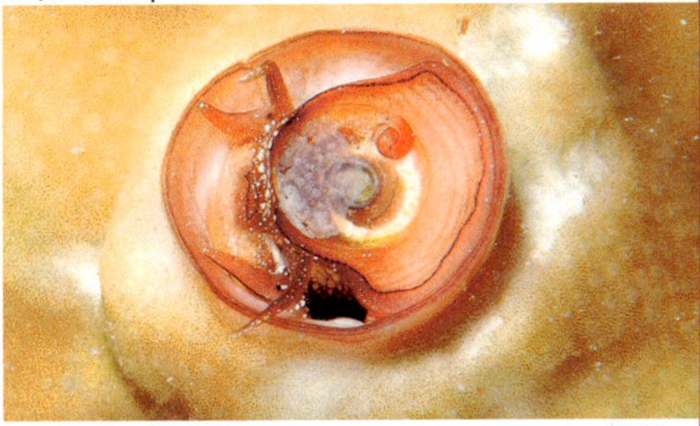

Keen's Worm Shell *Petaloconchus* cf. *keenae* WCP: Japan-Hawaii, 18 mm (diameter). ©JJ. ID tentative.

Daid's Worm Snail *Thylacodes* cf. *daidai* WCP: Japan, Philippines to Marshalls, 16 mm. ©JJ.

Mottled Worm Snail *Thylacodes* sp. Philippines, 18 mm. Animal grey with yellow mottling. ID tentative.

WORM SHELLS - VERMETIDAE

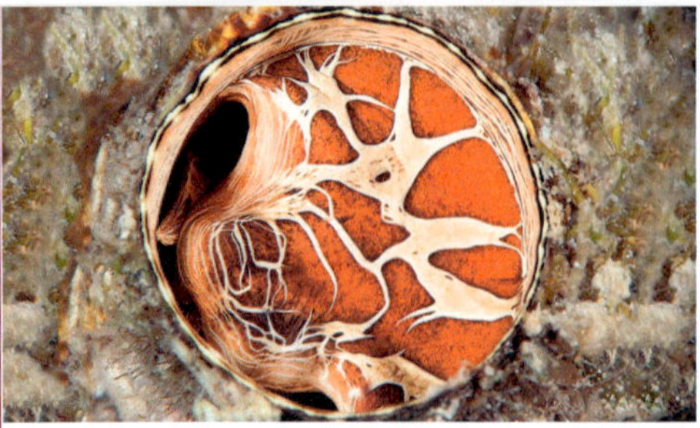

Grand Coral Worm Snail *Thylacodes grandis* WP: Philippines to Australia and New Guinea, 25 mm (diameter). Living animal of variable coloration. Photo on the right - mucus net in process of catching zooplankton.

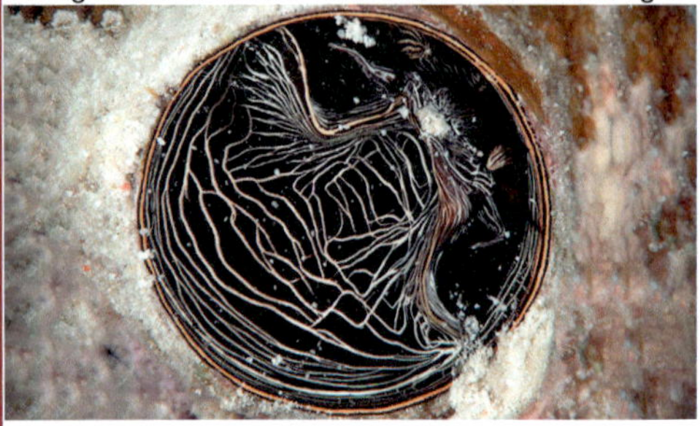

Grand Coral Worm Snail *Thylacodes grandis* (cont'd) Black color morph.

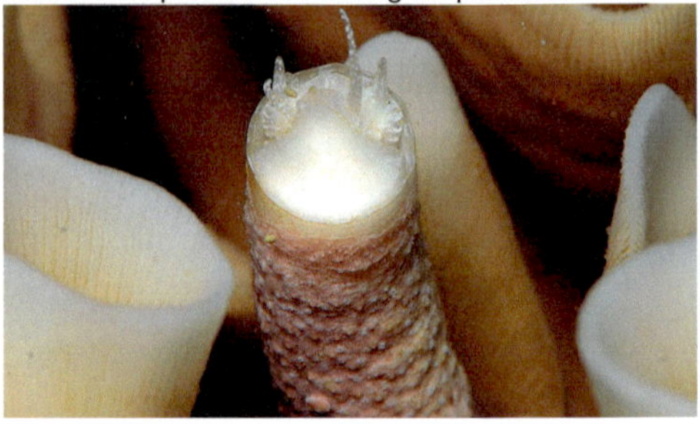

White Worm Snail *Thylacodes* sp. CP, Marshalls, 15 mm. Living animal white. ©JJ

Variable Worm Snail *Thylacodes variabilis* WCP, 15 mm. Living animal orange. ©JJ.

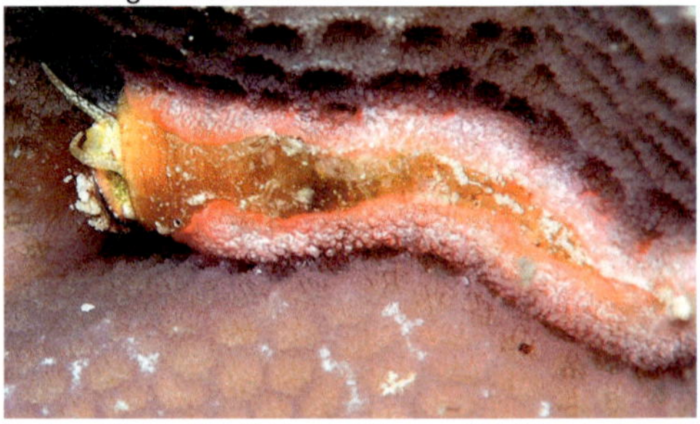

Yellow Worm Snail *Vermetidae* sp. CP, Marshalls, 15 mm. Living animal yellow. ©JJ. ID tentative.

Yellow-Tube Worm Snail *Vermetidae* sp. CP, 15 mm. ©JJ. ID tentative.

White-Shell Worm Snail *Vermetidae* sp. CP, Marshalls, 15 mm. ©JJ. ID tentative.

XENOPHORIDAE

Rough Carrier Shell *Xenophora cerea* IWP, 82 mm. Cements rubble to shell edges. © Uwe Weinreich.

Kono's Carrier Shell *Xenophora mekranensis konoi* IWP, 105 mm. Depth >90 m. © E.G de Suduiraut.

Impending Cerith *Bittium impendens* IP: S. Africa to Marshalls, 4 mm. Tiny conical shell. ©JJ.

Uneven Cerith *Cerithium salebrosum* IWP: Madagascar to New Caledonia, 37 mm.

Interrupted Cerith *Cerithium* cf. *alutaceum* WP: Philippines to Marshalls, 15 mm. Sharply conical shell, spiral sculpture. Variable coloration, often with interrupted dark bands. Translucent animal with white dots. ©JJ.

Black Lipped Cerith *Cerithium atromarginatum* IP: E. Africa to Japan and Hawaii, 15 mm. Outer lip with black stripes. Animal white with yellow pigment and red dotting. ©JJ.

Clavus-Like Cerith *Cerithium claviforme* WCP: Japan to Marshall Islands, 31 mm. Tan to red-brown shell, rows of spines with black tops, yellow, orange or mauve aperture. Pale animal with red pigment. ©JJ.

Spinose Cerith *Cerithium echinatum* IP: Red Sea, East Africa to Japan and Hawaii, 80 mm. Shell with nodulose spiral cords, cream with brown spots. Animal pinkish with yellow pigment. Right: © David Massemin.

Dirty Cerith *Cerithium egenum* IP: Red Sea, E. Africa to Japan and Marshalls, 12 mm. White shell with black or brown blotches. Animal white, with yellow dots and dark-brown spots. ©JJ

Strong Cerith *Cerithium munitum* WCP: Japan to New Caledonia and Marshall Islands, 50 mm. White or yellow shell, animal white with yellow pigment. ©JJ.

Island Horn *Cerithium nesioticum* IP: Red Sea, E. Africa to Marshalls, also Mediterranean. White shell with faint brown spotting, animal white or pale yellow-greenish. ©JJ.

Pacific Cerith *Cerithium pacificum* WCP: Philippines, Marshalls, 30 mm. Smooth fawn shell. ©SJ.

Beaked Cerith *Cerithium rostratum* IP: E. Africa to Hawaii, 27 mm. Common on intertidal grass beds.

Channeled Cerith *Cerithium punctatum* IP, Red Sea, East Africa to Japan and Marshalls, 20 mm. White shell with spirals of black spots. Animal pale grey with white dots and black spots.

Flat Sided Cerith *Cerithium balteatum* WP: Thailand to New Caledonia, 30 mm. NMHN-2018.

Zebra Cerith *Cerithium zebrum* IWP, 13 mm. White shell with brown spiral bands, animal white. ©JJ.

Purple Cerith *Cerithium* sp. CP, Marshalls, 10 mm. Elongated purple shell with white band. ©JJ

Brownish Cerith *Cerithium* sp. CP, Marshalls, 11 mm. Elongated white shell with brown areas. ©JJ

Pink Cerith *Cerithium* sp. CP, Marshall Islands, 15 mm. Shell pink with white spots. ©JJ. ID tentative.

Banded Cerith *Cerithium* sp. Pink shell, brown spiral bands. Probably form of the previous species ©JJ .

Traill's Cerith *Cerithium traillii* IWP: Madagascar to New Caledonia, 50 mm. Fawn shell with brown streaks.

Lavender Cerith *Cerithium* sp. CP, Marshall Islands, 12 mm. Pale lavender shell, white-dotted animal. ©JJ

Zoned Cerith *Cerithium zonatum* IWP: Madagascar to Japan and New Caledonia, 57 mm. Coloration variable, from white with spiral black bands to fawn, red brown or purplish-brown.

B&W Cerith *Clypeomorus* sp. CP, Marshall Islands, 18 mm. White shell with black streaks. ©JJ.

Koperberg's Cerith *Pictorium koperbergi* WCP, 9 mm. Small shell with pinkish lips. Animal purplish-red. ©JJ.

Aluco Vertagus *Pseudovertagus aluco* WCP, 10 cm. Shell with brown spotting, expanded end of outer lip.

Noble Creeper *Pseudovertagus nobilis* IWP: E. Africa Philippines, 172 mm. Big shell, fawn with pink hue.

White Cerith *Rhinoclavis fasciata* IP: Red Sea and Madagascar to New Caledonia and Marshall Islands, 95 mm. Slender shell (length/width ratio 4:1). Dark orange-brown oval operculum.

White Cerith *Rhinoclavis fasciata* (cont'd) Coloration variable, white to light brown, often with brown spiral bands or blotches. Like other family members, feed on algae and detritus, mostly below the sand surface.

White Cerith *Rhinoclavis fasciata* (cont'd) One of the most common species of Cerithiidae. Animal white: color form with white shell and longitudinal brown stripes (Marshall Islands) ©JJ.

Rough Vertagus *Rhinoclavis aspera* IP: Red Sea, East Africa to Hawaii and Polynesia, 60 mm. White to creamyellow shell, 3 rows of spines on each whorl, often with thin brown spiral bands. Animal body white.

Diadem Vertagus *Rhinoclavis diadema* IP: East Africa to Philippines and Marshall Islands, 40 mm. White shell with brown dots on spiral bands. Animal body white white with fine brown lines. ©JJ

Koch's Vertagus *Rhinoclavis* cf. *kochi* IWP: Red Sea and E. Africa to Philippines, 45 mm.

Long-Snouted Vertagus *Rhinoclavis longicaudata* IP: Mauritius to Fiji, 50 mm. ©JJ.

Obelisk Vertagus *Rhinoclavis sinensis* IP: Red Sea, East Africa to Hawaii and Polynesia, 70 mm. Sculpture mainly spiral, without prominent nodules. Shell white to fawn with brown spots. Body with red pigment. ©JJ.

Deceptive Royella *Royella sinon* IP: Red Sea, East Africa to Japan and Marshall Islands, 30 mm. White shell with prominent spiral rows of spines. Gap between spines mostly brown. Animal white with grey tentacles. ©JJ

Modulus Snail *Indomodulus tectum* IP: Red Sea, East Africa to New Caledonia and Polynesia, Small family, shells with unique aperture and siphonal area design. Animal pinkish with white spots. H. Zell (insertion), ©JJ.

Unsuitable Hinea *Hinea inepta* IP, 8 mm. Cream shell w/brown spiral bands. Animal body banded too. ©JJ.

Ribbed Clusterwink *Planaxis sulcatus* IWP, 35 mm. Intertidal rocks with wave action. ©SJ.

Varicose Wentletrap *Cirsotrema varicosum* IP: E. Africa to the Philippines, New Caledonia and Marshall Isl, 70 mm. Associated with the anemone *Heteractis crispa*. Feeds by insertion of its proboscis into a sea anemone. ©JJ.

Good Wetletrap *Cirsotrema bonum* IWP: Oman, Philippines to N. Caledonia, 40 mm.

Toothy Wentletrap *Cycloscala* cf. *sardellae* WP, 8 mm. Axial ribs with distinct crenulations. © S. Jazwinski.

Pea Wentletrap *Cycloscala peaseana* IP: Red Sea, East Africa to Tahiti and Marshalls, 11 mm. White shell with widely spaced axial lamellae. White animal with tan tentacles. ©JJ.

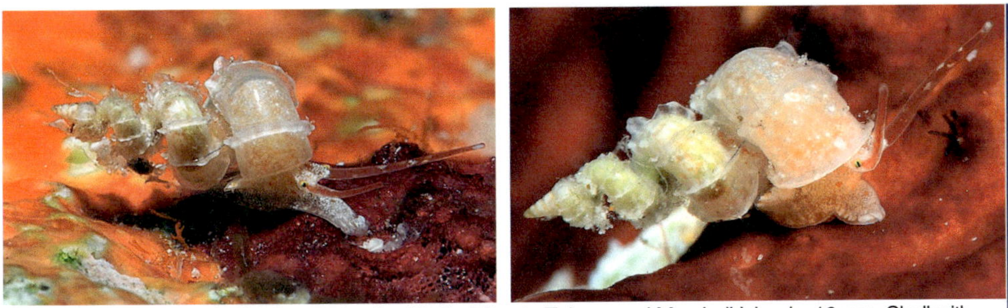

Transparent Wentletrap *Cycloscala hyalina* IP: Red Sea to Taiwan and Marshall Islands, 16 mm. Shell with widely spaced axial lamellae. Feeds on *Boloceroides* swimming anemones. © Christina Sylvester, ©JJ.

Violet Sea-Snail *Janthina janthina* Circumtropical, 40 mm. Lives pelagically, riding the underside of a bubble raft it creates. It feeds on the floating siphonophores like Portuguese man-o-war. © Christina Sylvester, ©JJ.

Twila's Wentletrap *Epifungium twilae* IP: Red Sea to Australia, Taiwan and Marshalls, 17 mm. Creamy white shell. Found with several fungiid corals host species. Clung with mucous threads to the underside of a coral. ©JJ.

Ulu Wentletrap *Epifungium ulu* IP: Red Sea, Maldives to Australia and Hawaii, 30 mm. Elongate-conical, creamy white shell. Found with several fungiid corals host species. ©JJ.

Golden-Spotted Wentletrap *Epitonium aureomaculatum* WP: Philippines, 34 mm. Creamy white with areas of brown pigment shell. Widely spaced axial lamellae. Animal white. ©JJ.

Gray's Wentletrap *Epitonium grayi* IP: E. Africa to Marshalls, 29 mm. White shell with closely spaced fine axial lamellae. Found under a sand near sea anemone *Stichodactyla haddoni*. ©JJ

Marble Wentletrap *Epitonium marmoratum* IP, 25 mm. White shell with cloudy brown markings. ©JJ.

Pyram Wentletrap *Epitonium pyramidale* IP, 50 mm. Preys on fire anemone *Actinodendron arboreum*. ©JJ.

Replicating Wentletrap *Epitonium replicatum* IP: Red Sea, South Africa to Marshall Islands, 18 mm. Shell with large widely spaced axial lamellae. Translucent animal with opaque white areas. ©JJ

Precious Wentletrap *Epitonium scalare* IWP, 72 mm. MNHN Koumac 2018. © D. Massemin.

White-Spotted Wentletrap *Epitonium* sp. Indonesia, 25 mm. On the picture: night attack on small anemone.

Red Wentletrap *Epitonium* sp. CP: Marshall Islands, 35 mm. Translucent-white shell, reddish animal with opaque white spots. Was found preying on the small anemone (photo on the right). ©SJ

Small Wentletrap *Epitonium* sp. CP, Marshalls, 20 mm. Head with yellow tentacles. © C. Sylvester.

White Wentletrap *Epitonium* sp. CP, 24 mm. Close to *E. irregulare*, but has more closely spaced ribs. ©JJ.

Golden Wentletrap *Epidendrium aureum* IP: Red Sea to Japan, Indonesia, Oceania, 15 mm. Bright yellow snails, they feed on *Tubastraea* tube corals, depositing egg masses on the corals (photo on the right).

E. aureum differ from *E. billeeanum* in having a dark purplish red protoconch instead of a yellow one ©JJ.

Small-Ribs Wentletrap *Surrepifungium costulatum* IP, Red Sea to Marshalls, 45 mm. On fungiid corals. ©JJ

Blue Miniature Cerith *Cerithiopsidae* sp. CP: Marshalls, 4 mm. Unidentified tiny snail. ©JJ.

Banded Miniature Cerith *Horologica* sp. CP: Marshalls, 5 mm. © Christina Sylvester. ID tentative.

Arduini Miniature Cerith *Joculator arduinii* WCP: New Caledonia, Philippines, Marshalls, 4 mm. ©JJ.

Grey Miniature Cerith *Joculator* sp. CP: Marshalls, 5 mm. ©JJ. ID tentative.

Mcmichael's Snail *Aclophoropsis mcmichaeli* WCP: Philippines to Australia and Marshalls, 5 mm. ©JJ.

Pale Minute Snail *Bouchetriphora pallida* WCP: Philippines to Australia and Hawaii, 8 mm.

Granulate Cautor *Cautor granulatus* IP: Madagascar to Australia and Marshall Islands, 15 mm. High-spire shell with spiral bands of rounded spines. White animal with dark-brown reticulation. ID tentative. ©JJ.

Hervier's Minute Snail *Cautotriphora hervieri* IP: Japan to Australia and Tuamotu, 15 mm. Brown to purple-brown and red shell. Like other family members, the shell with left-handed coiling. Animal red with yellow mottling. ©JJ

Grainy Coriophora *Coriophora granosa* IP: Red Sea, Madagascar to Polynesia, 8 mm. Animal red. ©JJ.

Grainy Coriophora *Coriophora* cf. *granosa* Marshalls, 6 mm. Similar to *C. granosa*, but body color differ. ©JJ.

Jazwinski's Minute Snail *Cautotriphora jazwinskii* CP: Marshall Islands, 7 mm. Red shell, white body. ©JJ.

Kosuge Euthymella *Euthymella kosugei* WCP: Australia to Marshall Islands, 15 mm. ©JJ.

Regal Euthymella *Euthymella regalis* WCP: Australia to Philippines and Marshals, 20 mm. ©JJ.

Reddish Euthymella *Euthymella* sp. CP: Marshalls, 10 mm. Red body with yellow spots. ©JJ.

Roughest Inella *Inella asperrima* WCP: Taiwan to N. Guinea and Marshalls, 20 mm. ©JJ.

Yellow-Bodied Inella *Inella* sp. CP: Marshalls, 15 mm. Grey to fawn shell, animal body bright yellow. ©JJ.

Ryo Suke's Inella *Inella ryosukei* WCP: Japan to Philippines and Marshall Islands, 14 mm. White to fawn shell, animal body white. ©JJ.

Golden-Orange Inella *Inella* sp. CP: Marshalls, 14 mm. Shell and animal body of orange color. ©JJ.

Resembling Iniforis *Iniforis aemulans* CP: Hawaii to Marshalls, 10 mm. ©JJ.

Pretty Iniforis *Iniforis formosula* WCP: Japan, Philippines to Marshalls, 15 mm. ©JJ.

Ikuko's Iniforis *Iniforis ikukoae* WCP: Japan to Marshall Islands, 11 mm. ©JJ.

Violet Iniforis *Iniforis violacea* WCP, 10 mm. Seem close to *I. violacea*, described as pure violet. ©JJ.

Violet Iniforis *Iniforis* cf. *violacea* Marshalls, 14 mm. Close to *I. violacea*, but have extra row of beads. ©JJ.

Mourning Mastonia *Mastonia funebris* WP: Australia, Marshalls, 6 mm. Brown and fawn spiral bands. ©JJ.

Lambert's Mastonia *Mastonia lamberti* IP: Reunion to Marshalls, 9 m. © Stan Jazwinski.

Red Mastonia *Mastonia rubra* IP: Red Sea, E. Africa to Marshalls, 11 mm. Mauve shell, often with spiral bands of brownish granules. Animal body translucent with pattern of opaque white areas. ©JJ.

Lavender Mastonia *Mastonia* sp. CP: Marshalls, 13 mm. Purple with spiral bands of golden granules. ©JJ.

Dirty Mastonia *Mastonia squalida* WCP: Japan to Marshalls, 7 mm. Grey shell with brownish areas. ©JJ.

Lucid Monophorus *Monophorus lucidulus* WCP: Philippines to Marshalls, 5 mm. ©JJ.

4-Spotted Monophorus *Monophorus quadrimaculatus* WCP, 4 mm. © C. Sylvester.

Wheat-Flour Nanaphora *Nanaphora triticea* WCP: Japan to Hawaii, 4 mm. ©JJ.

Collar Triphora *Triphora collaris* WCP to Marshalls, 9 mm. Fawn and white shell, white animal body. ©JJ.

Ribboned Triphora *Triphora taeniolata* WCP: Philippines to Australia and Marshalls, 12 mm. Fawn or redorange shell with black&white spiral bands. Yellow/orange animal with dark red-brown dots. ©JJ.

Red-Bodied Triphorid *Triphorid* sp. CP: Marshall Islands, 12 mm. Unidentified, probably undescribed species. ID tentative. ©JJ

Red-Bodied Triphorid *Triphorid* sp. CP: Marshall Islands, 12 mm. The shell is close to *Cautor granulatus*, but the animal color is very different. ©JJ.

White Triphorid *Triphorid* sp. CP: Marshalls, 12 mm. Close to *B. pallida* but it differs in its red animal color.

Creamy Triphorid *Triphorid* sp. CP: Marshalls, 9 mm. Unidentified, probably undescribed species. ©JJ. IT.

Purple Triphorid *Triphorid* sp. CP: Micronesia, 9 mm. Unidentified, probably undescribed species. ©JJ. IT.

Black Triphorid *Triphorid* sp. CP: Micronesia, 10 mm. Unidentified, probably undescribed species. ©JJ. IT.

Brown Triphorid *Triphorid* sp. IP: Micronesia, 10 mm. Probably undescribed species. © C. Sylvester.

Abbot's Viriola *Viriola abbotti* WCP: Philippines to Hawaii, 9 mm. Translucent-yellowish body. ©JJ.

Bayan's Viriola *Viriola bayani* WCP: Philippines to Micronesia, 25 mm. Body brown with white spots. ©JJ.

Reticulated Viriola *Viriola cancellata* IP: Red Sea, Reunion to Philippines and Micronesia, 15 mm. ©JJ.

Intergranular Viriola *Viriola intergranosa* IP: Reunion to Taiwan, New Caledonia and Micronesia, 7 mm. Shell and animal body of grey (or brown) colors with white spots/stripes. ©JJ.

Pagoda Snail *Viriola pagodus* WCP: Philippines to Micronesia, 50 mm. Light brown shell, spiral ribs plain, without tubercles. Animal body light brown with white dotting. ©JJ.

Whitish Nassa *Nassarius albescens* IWP: Madagascar to Japan and Philippines, 25 mm. Shell with numerous small granules, cream or fawn, some banded or speckled with brown. Decorated with *Cytaeis* sp. hydroids.

Vanuatu Nassa *Nassarius* sp. CP: Micronesia, 17 mm. Elongate, slender shell with prominent axial ribs, not overridden by spirals, 6-7 whorls. Animal body cream in color, without spotting. ©JJ.

Acute-Ribbed Nassa *Nassarius* cf. *acuticostus* WP, 19 mm. Broad shell with wide-spaced axial ribs.

Prickly Nassa *Nassarius cinctellus* WP: Philippines to N. Caledonia, 10 mm. Axial ribs with large spines.

Neat Nassa *Nassarius concinnus* IP: Reunion to Hawaii, 25 mm. Creamy shells, some with brown bands.

Burned Nassa *Nassarius crematus* IP: E. Africa to Hawaii, 4 cm. White to creamy with brown spots. ©JJ.

Cake Nassa *Nassarius coronatus* IWP: Red Sea to Solomons. Shell with prominent shoulder nodules, cream, often with brown spots between the shoulder nodules. Animal body translucent, with opaque white pigment.

Distorted Nassa *Nassarius distortus* IWP, 35 mm. White to cream shell with slender axial ribs, banded with greenish-brown. Bands usually confined to gaps between axial ribs. Animal body white with black speckling.

Channeled Nassa *Nassarius dorsatus* IWP, 45 mm. Grey to brown shell, often with faint banding.

Shining Nassa *Nassarius ecstilbus* IWP, 11 mm. © Philippe & Guido Poppe www.poppe-images.com

Gaudy Nassa *Nassarius gaudiosus* IP: E. Africa to Japan, Fiji and Vanuatu, 29 mm. Light brown to cream or white shell with fine spiral brown lines. Animal body white with black dots. Left: © C. Sylvester.

Glans Nassa *Nassarius glans* IP: East Africa to Japan, New Caledonia and Polynesia, 55 mm. White to cream shell with fine spiral brown lines. White animal body with black spots. Photo on the right: ©SJ.

Glans Nassa *Nassarius glans* (cont-d) Juveniles mostly white with small granules. ©JJ.

Rough Nassa *Nassarius hirtus* CP: Hawaii, Polynesia, 30 mm. Smooth shell with slender axial ribs. ©SJ.

Odd Nassa *Nassarius* cf. *abyssicolus* SP: N. Caledonia, 15 mm. Creamy animal with black spots. IT.

Gruner's Nassa *Nassarius gruneri* IWP, 20 mm. Creamy shell with rows of numerous pointed granules.

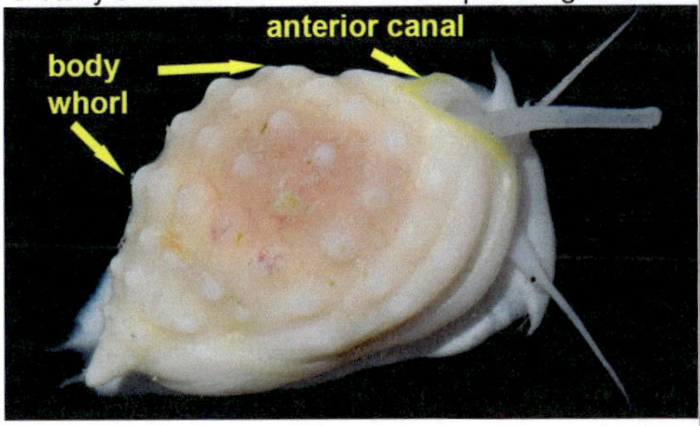

Granulated Nassa *Nassarius granifer* IWP, 23 mm. Shell with 4-5 spiral rows of big pointed granules on body whorl. Cream to light brown in color, anterior canal bordered with yellow. Animal body white.

Speckled Nassa *Nassarius limnaeiformis* WP, 23 mm. Variable in color, from dull brown to speckled fawn with fine red spiral bands. Animal body white with black spotting.

Rough Nassa *Nassarius horridus* IP: Madagascar, S. Africa to Marshalls, 15 mm. Outer lip with numerous irregular denticles. White to light brown, flecked with red-brown. Left: © C. Sylvester, right: ©JJ.

Common Nassa *Nassarius livescens* IWP, Madagascar to Japan and Indonesia, 30 mm. © R. Tan.

Margarite Nassa *Nassarius margaritifer* IWP, 35 mm. Cream shell with spiral lines, animal white and black.

Massemin's Nassa *Nassarius massemini* WP: Palau, Australia, N. Caledonia, 18 mm. © D. Massemin.

Multi-Ribbed Nassa *Nassarius multicostatus* WP, 15 mm. White to brownish shell with axial ribs.

Glossy Nassa *Nassarius nodiferus* WP: Singapore to New Zealand, 20 mm. ©JJ.

Olive Nassa *Nassarius olivaceus* IWP, 50 mm. Smooth olive to brown shell with spiral ridges. © Ria Tan.

Loaded Nassa *Nassarius oneratus* IP: Reunion to Japan and Marshall Islands, 18 mm. Cream to yellowish shell with spiral rows of nodules. Living animal creamy, covered by opaque white pigment. ©JJ.

Pimpled Nassa *Nassarius papillosus* IP: Reunion to Hawaii, 52 mm. Shell with spiral rows of rounded nodules. Cream in color, blotched or clouded with orange-brown. Photos: (left, right) ©SJ, ©JJ.

Black Nassa *Nassarius pullus* IWP, 25 mm. Shell with rounded axial ribs and impressed lines of interstices between them. Dorsal hump with dark brown blotch (highlighted by arrow). Right: © Ria Tan.

Reeveanu's Nassa *Nassarius reeveanus* IWP, 26 mm. Variable in color, shell from olive-brown to fawn with dark brown bands or mottled greenish with fine brown striation. Animal body dark grey, almost black.

Latticed Nassa *Nassarius shacklefordi* IP, 15 mm. Shell with 2 bands of a latticed pattern. © D. Massemin.

Sinu Sigeru Nassa *Nassarius sinusigerus* IWP, 12 mm. Shell w/brown spiral lines and gaps between nodules.

Splendid Nassa *Nassarius splendidulus* IP: Red Sea, Maldives to Japan and Hawaii, 25 mm. White shell with occasional red spotting and red protoconch. Live animal opaque white with black dots. ©JJ.

Comb Nassa *Nassarius troendleorum* WCP: Indonesia, Philippines to Micronesia, 25 mm. Outer lip with 5 prominent acute black teeth, bordered by red or brown pigment. Photos: (left, right)) ©JJ, © Bob Abela.

Lined Nassa *Nassarius teretiusculus* WP: S. China Sea, 17 mm. Pale yellow to creamy shell with fine purple spiral line. © Ria Tan.

Radiating Nassa *Nassarius radians* SP: Australia to Vanuatu, 19 mm. © Brian R. Mayes.

Textile Phos *Phos textilis* IWP: Mozambique to the Philippines, Indonesia and Australia, 15 mm.

Rosy Phos *Phos roseatus* IWP: Red Sea, E. Africa to the Philippines and Australia, 40 mm. Variable in color, orange, red-brown or purplish shell with occasional white spots. Animal body red-brown.

Sharp-Ribbed Phos *Phos muriculatus* WP: Japan to Indonesia and Australia, 35 mm.

Thorny Phos *Phos senticosus* IWP: Red Sea to N. Caledonia, 50 mm. Ribs with acute thorny spines.

Woven Pacific Phos *Phos textus* IWP: Seychelles to the Philippines and Australia, 30 mm. White to fawn shell with sharp axial ribs, animal body white to light brown with scattered black dots.

Blainville' Phos *Phos blainvillei* WCP: Philippines to Australia and Polynesia, 25 mm. ©JJ.

Round Nassa *Reticunassa rotunda* WCP, 7 mm. Shell with close set axial ribs. Animal with b&w dots. ©JJ.

Extended Nassa *Reticunassa neoproducta* IO: South and East Africa to Maldives, Sri Lanka and Bali, 13 mm. Dirty white to yellowish shell, often with brown banding. Animal body with white and black speckles.

Poppe' Nassa *Reticunassa poppeorum* WCP: Philippines and Indonesia to Marshall Islands, 12 mm. White to yellow or orange shell. Dorsal side of body whorl often with broad, orange-brown band in lower half (arrow!) ©JJ.

Headless Snail *Caducifer decapitatus* IP to Marshalls, 25 mm. Shell with broken off apex in adults. ©JJ.

Truncate Snail *Caducifer truncatus* IP to Marshalls, 25 mm. Shell with truncated apex in adults. ©JJ.

Strawberry Goblet *Clivipollia fragaria* IP: E. Africa to Philippines and Marshalls, 30 mm. Slender pink shell with pattern of dark bands with yellow or pale spots. Animal body reddish with white spotting. ©JJ.

Beautiful Goblet *Clivipollia pulchra* IP, 30 mm. Redbrown shell with darker spiral interspaces.©SJ.

Caroline's Engina *Engina carolinae* WP, 12 mm. Synonymized with *C. fragaria* (above) by some researchers.

Excellent Engina *Engina egregia* IP: S. Africa, Reunion, 26 mm. Slender shell with curved siphonal canal, light brown with darker axial ribs. Animal yellow-brownish wit black and white speckling. ©JJ.

Powdery Engina *Engina farinosa* IP: Reunion to Hawaii, 20 mm. White to creamy shell with prominent axial ribs. Animal body white with light brown speckling. ©JJ.

Lined Engina *Engina lineata* IP: East Africa to the Philippines and Micronesia. Light yellowish shell with black spiral bands and two distinct black spots on dorsal. Living animal striped. ©JJ.

Striped Engina *Engina mendicaria* NWP, 20 mm. Yellowish biconical shell with large black spiral bands. Living animal mostly black, with yellowish areas on foot. ©JJ.

Spike Engina *Engina spica* IP, 13 mm. White biconical shell with a pattern of interrupted brown spiral bands. Living animal white with brown spots. Photo on the right: ©JJ.

Banded Engina *Engina zonalis* WCP: Indonesia, Taiwan to Micronesia, 20 mm. White shell with spiral rows of prominent nodules and black bands between them. Animal body black with white speckles. ©JJ.

Flame Pisania *Pisania ignea* IP: Red Sea, East Africa to the Philippines and Micronesia, 52 mm. Slender redbrown shell with brown and white bands. Red animal with white reticulation. ©JJ.

Mottled Pisania *Pisania* sp. WP, 24 mm. Shell with pattern of brown and white, animal with brown dots. ©JJ

Checkered Pisania *Pisania* sp. WP, 20 mm. Yellowish shell with brown spiral lines and axial stripes.

Sugimoto's Pisania *Pisania* cf. *sugimotoi* WP, 22 mm. Pinkish shell with brown spiral bands.

Wagner's Goblet *Pollia wagneri* IP: Reunion to Marshalls, 35 mm. Fine brown bands and large spots. ©SJ.

Waved Goblet *Pollia undosa* IWP, 45 mm. White to orange shell with dark-brown spiral bands and orange aperture margin. Living animal tan with orange hue and black speckles.

Pink Prodotia *Prodotia iostoma* IP: S & E. Africa to Hawaii, 36 mm. Slender pink shell with spiral rows of pale or yellow granules. Animal white with brown spots. ©JJ.

Gold-Banded Latirus *Crassicantharus noumeensis* IP: Reunion to Hawaii, 25 mm. ©JJ.

Sowerby' Tulip Snail *Benimakia sowerbyi* IP: Red Sea to Micronesia, 30 mm. ©JJ.

Wavy Spindle *Cyrtulus undatus* IWP, 253 mm. Fusiform shell with well developed siphonal canal.

Precious Stone Shell *Latirolagena smaragdulus* IP: Red Sea to CP, 6 cm. Rounded, not nodulose shell.

Spear Latirus *Dolicholatirus lancea* IP: Red Sea to Micronesia, 57 mm. Extremely slender, spear-shaped shell with orange spiral bands. Live animal white with orange reticulation. ©JJ.

Filamentous Horse Conch *Filifusus filamentosus* IP: Red Sea, East Africa to Philippines and Melanesia, 180 mm. Orange-brown fusiform shell with pointed spire. Live animal dark red with white speckles. ©SJ.

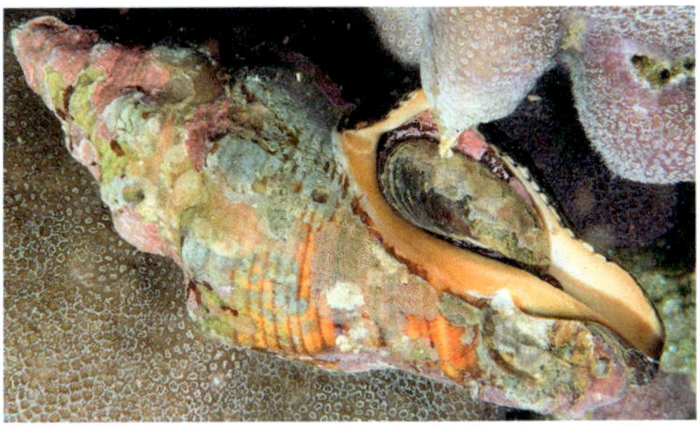

Gibbose Latirus *Latirus gibbulus* IWP, 117 mm. Fawn shell with nodules on the shoulders, dark spiral bands.

Pale Spindle *Latirus* sp. CP: Micronesia, 14 mm. Pale shell with prominent axial ribs. ©JJ.

Spotted Latirus *Latirus maculatus* IP: Mauritius to the Philippines and Polynesia. Light-brown to pink-brown shell with large brown spots. Live animal red with white dots. ©JJ.

Nicobar Spindle *Marmorofusus nicobaricus* IP: Thailand to Hawaii,180 mm. Shell with long siphonal canal and brown marks. Adults with strong shoulder nodules. Animal white with grey speckles.

Knobbed Spindle *Nodolatirus nodatus* IP: Mauritius to Micronesia, 100 mm. Fusiform knobbed shell with purplish aperture, live animal bright red with white speckles. ©JJ.

Spotted Spindle *Peristernia chlorostoma* IP, 21 mm. White or creamy shell with dark brown spots dorsally, orange aperture. Live animal bright red with white speckles, operculum orange. ©JJ.

Fine-Net Peristernia *Peristernia nassatula* IP: Red Sea, E. Africa to Micronesia, 45 mm. White shell with pinkreddish or brown spiral bands. Animal red with white dots, operculum black. ©JJ.

Flesh-Colored Peristernia *Peristernia reincarnata* IP: Red Sea to the Philippines, Australia and Micronesia, 34 mm. Orange or fawn axial ribs, brown interstices, purple aperture. Animal pink, foot with orange margin. ©JJ.

Singed Peristernia *Peristernia ustulata* IWP, 35 mm. Cream to red-brown shell, pink animal.

Trapezium Horse Conch *Pleuroploca trapezium* IP, 280 mm. Angulate shoulder, prominent nodules. ©SJ.

Ribbed Turrilatirus *Turrilatirus craticulatus* IP: East Africa to Micronesia, 61 mm. Yellow shell with red-brown axial ribs. Live animal red with pale reticulation, foot orange. ©JJ.

Tower Latirus *Turrilatirus turritus* IP: Red Sea, E. Africa to Micronesia, 65 mm. Shell with broad axial ribs lacking nodes, orange with black spiral cords. Animal red with fine white streaks. ©SJ.

Cuming's Dwarf Triton *Colubraria cumingi* IP, 50 mm. Feeds parasitizing on sleeping fish. ©JJ.

Tender Dwarf Triton *Colubraria tenera* IP: E. Africa to Micronesia, 85 mm. Sucks the blood of fish. ©SJ.

Maculated Dwarf Triton *Colubraria muricata* IP: E. Africa to Polynesia, Japan to Australia,115 mm. Shell with high spire, shot aperture. Mostly nocturnal, parasitic upon sleeping fish, usually parrotfishes. ©SJ.

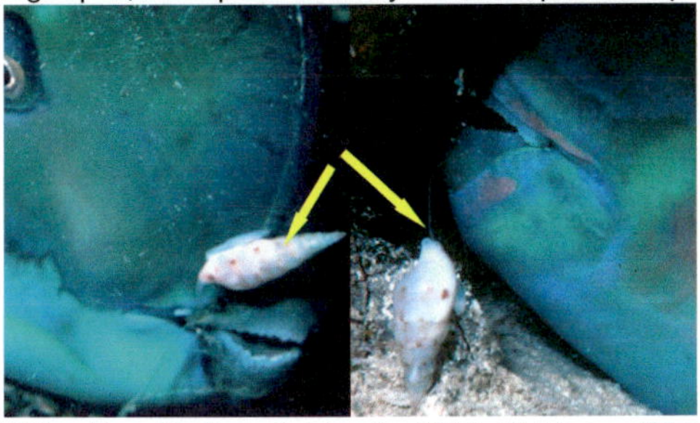

Shiny Dwarf Triton *Colubraria nitidula* IP, 66 mm. White smooth glossy shell with some brown streaks. Animal white, yellow operculum. Mostly nocturnal, parasitic upon sleeping fish. ©JJ.

Obscure Dwarf Triton *Colubraria obscura* IP, also Atlantic, 66 mm. Fawn to pale brown shell with pattern of brown interrupted spiral bands. Mostly nocturnal, parasitic upon sleeping fish. ©JJ.

Twisted Dwarf Triton *Colubraria tortuosa* IP, 50 mm. Brown shell with a pattern of zigzag brown spots. Animal white, with black siphon. Nocturnal, parasitic upon sleeping fish. Photos: (left, right) © S. Jazwinski, ©JJ.

Slender Dove Shell *Aesopus clausiliformis* IP: Reunion to the Philippines and Micronesia, 15 mm. Smooth shell, variable in color, pinkish and fawn to mauve and red. White animal with brown head and tentacles. ©JJ.

Deshayes's Dove Shell *Euplica deshayesii* WCP: Philippines to Australia and Micronesia, 15 mm. Globose pale yellow or fawn shell, live animal foot with red margin head and siphon pink or red. ©JJ.

Minute Dove Shell *Euplica ionida* IP: E. Africa to Japan and Micronesia, 7 mm. Tiny shell with axial ribs, extremely variable in color: from dark red to greenish, pale pink. Animal color matches the shell. ©JJ.

Livid Dove Shell *Euplica livescens* CP: Micronesia to Hawaii, 12 mm. Smooth red-brown shell. ©JJ.

Varicolored Dove Shell *Euplica scripta* IP, 22 mm. Shell with recognizable form, variable in color.

Orange Dove Shell *Euplica* sp. WP: Indonesia, 16 mm. Orange shell with faint brown bands, animal orange.

Globose Dove Shell *Euplica* sp. WP: Philippines, 15 mm. Tan globose shell, animal green with white dots.

Variable Dove *Shell* *Euplica varians* IP: Red Sea, E. Africa to Hawaii, 11 mm. Biconical shell with a series of large shoulder nodules. White to fawn and beige with wavy lines, brown blotches. Animal mottled brown. ©JJ.

Turtle Dove Shell *Euplica turturina* IP: East Africa to the Philippines and Hawaii,16 mm. Variable in color, adults globose, almost spherical, juveniles conical. Aperture with purple inner and outer lips and white denticles. ©JJ.

Turtle Dove Shell *Euplica turturina* (continued) Live animal color usually matches the shell. Photo on the right: pink juvenile (©JJ).

Diverse Dove Shell *Graphicomassa ligula* IP: Red Sea, East Africa to New Zealand and Micronesia, 25 mm. Biconic thick shell, white, with variably colored mottling and pattern of spiral bands. ©JJ.

Diverse Dove Shell *Graphicomassa ligula* (continued) this form was previously treated as a separate species, Mitrella albina ©JJ.

Diverse Dove Shell *Graphicomassa ligula* (continued) color forms: dark grey pattern and pale-brown netpattern. Animal translucent with opaque white pigment, anterior foot with brownish bands. Right photo ©JJ.

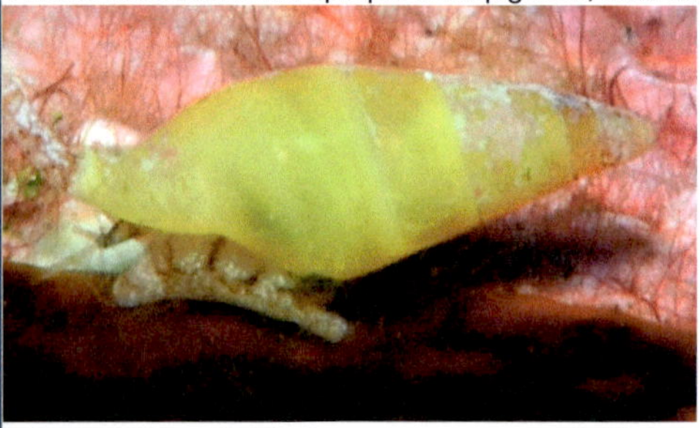

Obscure Dove Shell *Graphicomassa ligula* (continued) Color forms, possibly a separate species, yellow and with light brown bands. ©JJ.

Marquesas Dove Shell *Metanachis marquesa* IP: Reunion to Malaysia and Marquesas, 15 mm. Slender shell, variable in color. Some with dark-brown spiral bands. Thick pale foot with brown spotting. ©JJ.

Sweetheart Dove Shell *Indomitrella puella* WCP, 24 mm. White brown-spotted body. © D. Massemin.

Hanley's Dove Shell *Mitrella floccata hanleyi* IWP: Reunion to Fiji, 9 mm.

Bedewed Dove Shell *Mitrella rorida* IP, Reunion to Micronesia, 10 mm. Smooth, glossy shell. ©JJ.

Brown-Marked Dove Shell *Mitrella* sp.17 Micronesia, 9 mm. Shell and body with brown pattern. ©JJ.

Banded Dove Shell *Mitrella* sp. CP: Micronesia, 10 mm. Biconic pale-brown shells with black@white spiral bands. Appears to be undescribed species, close to *Mitrella loyaltyensis* ©JJ.

White-Banded Dove Shell *Mokumea divaricata* IP: E. Africa to Australia, Japan, Micronesia, 5 mm. Slender yellow-brown shell with brown spiral bands and white whorl borders. White animal with black pigment. ©JJ.

Marble Dove Shell *Pardalinops marmorata* WCP: Australia to Philippines and Micronesia, 12 mm. Orange-brown on white shells, some with white spots or patches. Animal foot and siphon with yellow ends. Right photo: ©JJ.

Tortoise Dove Shell *Pardalinops testudinaria* IWP: Red Sea, East Africa, the Philippines to Australia, 41 mm. Tan or brown shell with pattern of dark spots, or white spots on dark field. Animal with grey or brown pigment.

Beautiful Dove Shell *Pseudamycla* aff. *formosa* WP: Japan to Philippines, 11 mm.

Obscure Dove Shell *Pyrene obscura* WP, 25 mm. White shell with pattern of wavy black lines.

Yellow Dove Shell *Pyrene flava* IP: E. Africa to Micronesia, 30 mm. Fawn shell, usually with white spots, often a dark brown band below suture. Animal body red-brown with patches of opaque white. ©JJ.

Striated Dove Shell *Pyrene* sp. CP, Marshalls, 12 mm. Shell with pattern of dark blotches and fine lines. ©JJ.

Punctate Dove Shell *Pyrene punctata* IP: E. Africa, Japan to Australia, 26 mm. © E.G de Suduiraut.

Grey Dove Shell *Pyrene* sp. WP: Indonesia, 14 mm. Grey with pattern of black and white spots. ID tentative.

Half-Decorated Dove Shell *Pyreneola semipicta* WP: Indonesia, 9 mm. Shell with pattern of brown dashes.

Melvill's Dove Shell *Pyreneola melvilli* WCP: Indonesia, Philippines to Marshalls, 5 mm. White shell striped by short narrow, brown lines. ©JJ. Photo on the right - juvenile, about 3 mm.

Weak Dove Shell *Zafra* cf. *debilis* CP: Marshalls, 3 mm. Possibly undescribed species. Morphology reminds Zafra debilis. Biconical pale shell, axial ribs with yellow interstices. Animal white with black pigment. ©JJ.

Ocellate Dove Shell *Zafra ocellatula* IP: Reunion to Marshalls, 6 mm. Body whorl with ocellate pattern. ©JJ.

Small Dove Shell *Zafra* sp. CP: Micronesia, 4 mm. Shell with pattern of brown lines and white bands. ©JJ.

Dark-Skirt Dove Snail *Zafra* sp. CP: Micronesia, 4 mm. Axial ribs, dark area on body whorl. ©JJ.

Grey Dove Snail *Zafra* sp. CP: Micronesia, 5 mm. White shell with grey bands. ©JJ.

Brown Dove Snail *Zafra* sp. CP: Micronesia, 6 mm. Dark-grey to red brown shell with red protoconch. Animal body translucent with white dots. ©JJ.

Brazier Dove Snail *Zafra vercoi* IP: Red Sea, E. Africa to Micronesia, 4 mm. ©JJ.

Pleurifera-Like Dove Shell *Zafrona pleuriferoides* 6 mm. CP: Micronesia to Society, 6 mm. ©JJ.

Nebulous Dove Shell *Zafrona isomella* var. *nebulosa* WCP: China Sea to Hawaii, N. Guinea, Fiji, 8 mm. Slender white shell with pattern of brown spiral bands, often limited with patches of brown background. ©JJ.

Bailey Cone *Conasprella baileyi* IWP, 32 mm. MNHN Kanacono 2016. © David Massemin.

Hopwood's Cone *Conasprella hopwoodi* WP, 32 mm. © Philippe & Guido Poppe www.poppe-images.com.

Tirard's Cone *Conasprella tirardi* SP, 31 mm. MNHN Koumac 2018. © David Massemin.

Guilded Cone *Conus aulicus* IP: Madagascar to Japan, F. Polynesia, 163 mm. Preys on gastropods.

Sharp-Angled Cone *Conus acutangulus* IWP, 46 mm. Shell with high spire and strong spiral ribbons. Mostly brown with white spots. Live animal white, patterned with brown.

Admiral Cone *Conus ammiralis* IP: Red Sea to Marshall Islands, 109 mm. White shell with a complex pattern of brown spiral bands and white tents. Animal white, patterned with black and brown. Right photo: ©JJ.

Sand-Dusted Cone *Conus arenatus* IP: Red Sea, E. Africa to Tuamotu, 90 mm. White with spiral rows of dark brown dots. Foot cream, mottled with brown. Preys on worms: eunicids, nereids, capitellids. Right photo: ©JJ.

Gilded Cone *Conus auratinus* WCP, 120 mm. Foot white with brown blotches and lines. ©SJ.

Gold-Leaf Cone *Conus auricomus* IP: E. Africa to Japan, Hawaii and F. Polynesia, 69 mm. ©JJ.

Gold-Leaf Cone *Conus auricomus* (cont'd) Shell with high spire, white to pale yellow, patterned with brown and numerous white or yellow tents. Foot white with brown dots. ©JJ.

Golden Cone *Conus aureus* IP: E. Africa to Maldives, Japan and Tuamotu, 80 mm. White to pale yellow with a network of fine brown lines and brown blotches. ©JJ. Photo on the right: juvenile,15 mm.

Banded Marble Cone *Conus bandanus* IP: E. Africa to Philippines and Hawaii, 150 mm. ©SJ.

Captain Cone *Conus capitaneus* IP: E. Africa to Hawaii, Japan and Australia, 98 mm. ©JJ.

Captain Cone *Conus capitaneus* (cont'd) Shell with a low spire and 2 pale yellow bands: on the shoulder and in the centre. Animal greenish yellow dotted with black. Preys on gastropods. Photos: (left, right) © B. Abela, ©JJ.

Boetian Cone *Conus boeticus* IWP: E. Africa to Japan and Fiji, 40 mm. Yellowish low-spire shell with blackish brown markings. Live animal yellowish with brown mottling. Photos: (left, right) ©JJ, © Bob Abela.

Bubble Cone *Conus bullatus* IP: E. Africa to Marquesas and Hawaii, 82 mm. Shell variable in color: orange to fawn, with fine white axial lines. Foot white with brown lines and blotches. Photos: (left, right) ©SJ, © Bob Abela.

Cathy's Cone *Conus cathyae* WCP, 135 mm. Reddish brown shell with small white tents. © B. Abela.

Vermiculate Cone *Conus chaldaeus* IP: E. Africa to Japan and Central America, 59 mm. ©JJ.

Cat Cone *Conus catus* IP: E. Africa to Japan and Hawaii, 52 mm. Low-spire shell, variable in color: from yellow-brown to dark brown or red, some with brown clouds. Foot white with brown pigment. Photo: (left, right) ©SJ, ©JJ.

Celine's Cone *Conus coelinae* WCP: Australia, N. Caledonia to Hawaii, 128 mm. Shell yellowish, covered by brown periostracum. Foot brown with black spot anterior. Photo: (left, right) ©JJ, © D. Massemin.

Cropped Cone *Conus circumcisus* WCP: Philippines to Australia, Fiji and Marshalls, 100 mm. Shell yellowish orange or yellow, some with dark brown blotches. Preys on fishes. ©SJ.

Brazier Cone *Conus brazieri* WCP, 100 mm. Pale shell, without dark brown spots. Deep species, usually below 35 m. Was synonymized with C. circumcisus by some experts. Species delimitation awaits DNA analysis. ©SJ.

Sunburnt Cone *Conus cinereus* WP: Japan to Indonesia and Vanuatu, 57 mm. Beige foot spotted with black.

Singed Cone *Conus consors* IP: E. Africa to Marshalls, 118 mm. Yellow, 2 orange bands. © R. Groeneveld.

Coffee Cone *Conus coffeae* WCP: Philippines, Indonesia to Marshalls and F. Polynesia, 51 mm. White shell with violet aperture. Periostracum yellow to brown. Animal body red with white dots. Photo on the right: ©JJ.

Papal Cone *Pygmaeconus papalis* IP: Red Sea, E. Africa to Hawaii, 47 mm. Formerly *C. coronatus*. Grey to beige, spiral rows of alternating white and dark dots. Foot reddish with dark brown mottling. Right photo: ©JJ.

Saffron Cone *Conus crocatus* IP: E. Africa, Thailand to Marshalls and Samoa, 82 mm. Shell with rounded shoulder, orange with white or yellow tents. Foot yellow or orange, with black spots. ©JJ.

Cylindrical Cone *Conus cylindraceus* IP: E. Africa to Hawaii and Marshalls, 59 mm. Ovate shell with high spire and indistinct shoulder. White with pattern of brown or purple-brown. Photos: (left, right) ©SJ, ©JJ.

Distant Cone *Conus distans* IP: Red Sea, E. Africa to Hawaii and F. Polynesia, 150 mm. Shell white or yellowish. Periostracum dark brown, thick. Preys on eunicids. Photos (left, right - 1.5 cm. juvenile): ©SJ, ©JJ.

Coral Cone *Conus corallinus* WCP: Japan to Papua N. Guinea and Marshalls, 38 mm. ©SJ.

Ivory Cone *Conus eburneus* IP: E. Africa to F. Polynesia, 79 mm. Preys on eunicids, capitellids.

Ivory Cone *Conus eburneus* (cont'd) White shell with low spire and spiral rows of brown to black squarish spots. White foot with black spots. Photos (left, right - albino form): ©SJ, ©JJ.

Hebrew Cone *Conus ebraeus* IP: E. Africa to Japan, Australia, C. America, 62 mm. White shell with 3-4 spiral rows of black elongate squarish blotches. Black foot with beige margin. Preys on eunicids, nereids. ©JJ.

Magnificent Cone *Conus episcopatus* IP: E. Africa to Japan, Marshalls, F. Polynesia, 115 mm. White shell with light tent-like markings on the brown zones. White foot mottled with brown, black blotch in anterior part. ©JJ.

Cabrit's Cone *Conus exiguus* WP: Samoa to New Caledonia, Vietnam, 54 mm. © Yves Trenevet.

Yellow Pacific Cone *Conus flavidus* IP, 75 mm. Yellow shell with thick brownish periostracum. ©JJ.

Snowflake Cone *Conus floccatus* WCP: Indonesia to Philippines and Marshalls, 86 mm. White shell with numerous brown blotches, dots and axial lines. Animal white, mottled with brown, black blotch anteriorly. ©JJ.

Flowery Cone *Conus floridulus* IWP: Reunion to Maldives, Japan and N. Caledonia, 59 mm.

Grainy Cone *Conus granum* IWP, 40 mm. © Philippe & Guido Poppe www.poppe-images.com.

Geography Cone *Conus geographus* IP: Red Sea, East Africa to Japan and F. Polynesia, 166 mm. White with network of red-brown lines. Foot white with brown in the middle. Preys on fishes. Photo on the left: ©JJ.

Acorn Cone *Conus glans* IP: East Africa to Japan and F. Polynesia, 65 mm. White to pale blue shell with rounded shoulder and closely spaced, often granulose spiral ribs. Red foot with white dots. Preys on eunicids. ©JJ.

Glory of the Sea Cone *Conus gloriamaris* WCP: Indonesia to the Philippines and Solomons, 162 mm. ©SJ.

Lienard's Cone *Conus lienardi* WP: Philippines to N. Caledonia and Vanuatu, 63 mm. © D. Massemin.

Imperial Cone *Conus imperialis* IP: East Africa to Hawaii, 110 mm. Shell with angulate shoulder, white with 2 brown bands, split into streaks and blotches. Pink foot with brown mottling. Photos: (left, right) ©SJ, ©JJ.

Ambassador Cone *Conus legatus* IP: E. Africa to Japan and F. Polynesia, 63 mm. Pinkish-white shell with a network of fine brown lines and axial raws of white tents. Foot white with brown mottling and black spots. ©JJ.

Leopard Cone *Conus leopardus* IP: E. Africa to Hawaii, 222 mm. Shell white with spiral rows of dark brown spots. Periostracum yellowish or reddish brown. Preys on enteropneusts (acorn worms). ©SJ.

Lithograph Cone *Conus litoglyphus* IP: E. Africa to Hawaii, 75 mm. Brown shell with interrupted white bands at shoulder and below centre. Animal olive, mottled with dark brown. Preys on eunicids. ©JJ.

Lettered Cone *Conus litteratus* IP: E. Africa to Japan and Hawaii, 170 mm. White shell with 3 yellowish bands and spiral rows of dark-brown rounded spots. Animal pink, speckled with dark brown. Preys on capitellids. ©JJ.

Livid Cone *Conus lividus* IP: Red Sea, East Africa to the Philippines and Hawaii, 81 mm. Yellow shell with white bands at shoulder and below centre, aperture violet, apex pink. Preys on acorn worms, spaghetti worms. ©JJ.

Livid Cone *Conus lividus* (cont'd) Animal red with black speckles and white dots.

Yellow Cone *Conus luteus* WCP: Philippines to F. Polynesia and Marshalls, 54 mm. © S. Jazwinski.

Weasel Cone *Conus mustelinus* 107 mm, IP: Maldives to Marshalls. Yellow shell with 2 white bands. ©JJ.

Mariel's Cone *Conus marielae* CP: Marshall Islands to Marquesas and Tuamotus, 60 mm. ©SJ.

Magnificent Cone *Conus magnificus* IP: East Africa to F. Polynesia, 114 mm. Similar to *C. episcopatus*, differs in having a finer reticulate pattern, usually higher spire and a larger proportion of tent-like markings. ©JJ.

Magical Cone *Conus magus* WCP: Indonesia to Fiji, Marshalls and Japan, 94 mm. White to yellow, low-spire shell, white to yellow, with brown blotches, often dotted spiral lines. Periostracum brown, aperture white. ©JJ.

Magical Cone *Conus magus* (cont'd) Animal white, mottled with brown. Preys on fishes.

McBride' Cone *Conus mcbridei* WCP: Philippines, Hawaii, 11 mm. White shell, anterior 1/3 dark brown. ©JJ.

Marble Cone *Conus marmoreus* IP: India to Fiji and Marshalls, 150 mm. Shell with a pattern of white or yellow tent-like marks on black background. Photos (left, right - preying on juvenile cowrie) : ©SJ, ©JJ.

Soldier Cone *Conus miles* IP: E. Africa to Hawaii, 136 mm. White shell with 2 dark brown bands, fine brown axial lines and large brown patches. Animal dark grey or black. Preys on eunicids. Photo on the right: ©JJ.

Thousand-Spot Cone *Conus miliaris* IP: Red Sea, E. Africa to F. Polynesia and Marshalls, 43 mm. Pale-apricot shell with white zig-zag lines and spiral rows of brown and white dots. Animal yellow. Preys on eunicids. ©JJ.

Molucca Cone *Conus moluccensis* IP: Reunion to New Caledonia and Marshalls, 60 mm. © D. Massemin.

Minute Cone *Conus nanus* Philippines, Hawaii, 34 mm © Philippe & Guido Poppe www.poppe-images.com.

Morelet's Cone *Conus moreleti* IP: E. Africa, Maldives to F. Polynesia and Hawaii, 61 mm. Dark-brown shell, sometimes with a lighter central band. Dark-brown foot with red margin. Preys on terebellids. ©JJ.

Music Cone *Conus musicus* IP: Maldives to Australia, Japan and Marshalls, 30 mm. White or grey shell with a spiral rows of dots and dashes. Animal red with white dashes. Preys on nereids. Photos: (left, right) ©SJ, ©JJ.

Nut Cone *Conus nucleus* IP: Madagascar to Maldives, Japan to Marshalls, 25 mm. Orange to olive-brown shell with lighter central band and pale violet aperture. Animal yellow to orange. Photos: (left, right) ©JJ, ©SJ.

Nussatella Cone *Conus nussatella* IP: Red Sea, East Africa to Hawaii, 95 mm. White or yellow shell with spiral rows of small dark brown spots. Animal white with dark-brown mottling. ©JJ.

Obscure Cone *Conus obscurus* IP: E. Africa to Hawaii and F. Polynesia, 44 mm. Violet-brown shell with brown blotches and spiral rows of alternating brown-grey dots&dashes. Preys on fishes. ©JJ.

Omaria Cone *Conus omaria* IP: E. Africa to Japan, Marshalls and N. Caledonia, 86 mm. Pale yellow or cream shell with pattern of tent-like marking. White foot with red-brown streaks and black spots. Left: © Bob Abela.

Prefect Cone *Conus ochroleucus* WP, 88 mm. © Philippe & Guido Poppe www.poppe-images.com.

Parian Cone *Conus parius* WP, 46 mm. White shell with brown periostracum. © Rokus Groeneveld.

Feathered Cone *Conus pennaceus* IP: Indian Ocean from E. Africa to Maldives, Hawaii, 88 mm. Shell with a fine network of brown lines and tents. Animal reddish-brown with white mottling. Preys on gastropods. ©SJ.

Pertusus Cone *Conus pertusus* IWP, 69 mm. Pinkish shell with red clouds, dotted and dashed spiral lines. Animal yellow with dark brown dots. ©JJ.

Pertusus Cone *Conus pertusus* (cont'd) Dotted and dashed spiral lines clearly visible. ©JJ.

Oak Cone *Conus quercinus* IP, 140 mm. Yellow shell, brown spiral lines. Animal grey with brown dots. ©JJ.

Ringed Cone *Conus planorbis* IP: East Africa, Japan to Hawaii and F. Polynesia, 82 mm. Shell brown or redbrown with 2 faint pale spiral bands. Animal yellow with black spots. Photos: (left, right) © Bob Abela, ©JJ.

Flea-Bite Cone *Conus pulicarius* CWP: Indonesia to F. Polynesia and Hawaii, 75 mm. White shell with irregularly scattered small black spots. Preys on capitellids (worms). Photos: (left, right) SJ ©JJ.

Radiant Cone *Conus radiatus* WP, 109 mm. © Philippe & Guido Poppe www.poppe-images.com.

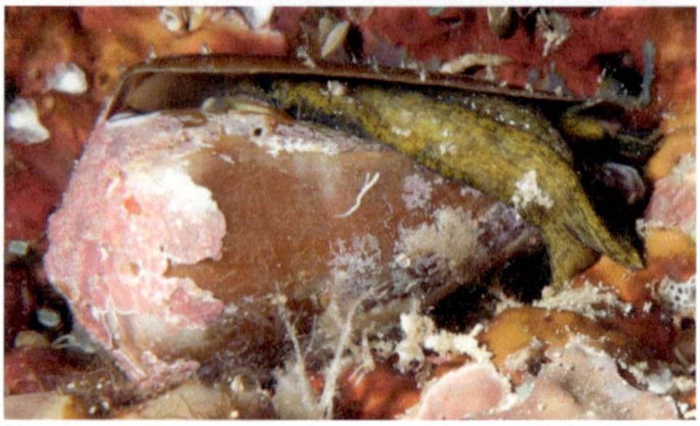

Rat Cone *Conus rattus* IP, 70 mm. Pale bluish shell with yellow to olive periostracum, animal yellow.

Netted Cone *Conus retifer* IP: E. Africa to Hawaii and F. Polynesia, 69 mm. Pale pink shell with rounded shoulder and pattern of black axial lines, fine brown lines and tent-like marks. Photos: (left, right) ©JJ , ©SJ.

Bloodstained Cone *Conus sanguinolentus* IP: E. Africa, Thailand to Polynesia, 65 mm. ©JJ.

Solomons Cone *Conus solomonensis* WP: Solomons, N. Guinea, 40 mm. Pale bluish with brown spots. ©SJ.

Marriage Cone *Conus sponsalis* IP: E. Africa to Hawaii and F. Polynesia, 34 mm. White shell with red-brown axial blurred bands. Animal pink with fine white stripes. Preys on worms: nereids, eunicids. ©JJ.

Fly-Specked Cone *Conus stercusmuscarum* WP, 64 mm. Pale grey or brown with spiral rows of brown dots and white dashes. Animal white, mottled with brown. Photo on the left ©SJ. Preys on fishes.

Striate Cone *Conus striatus* IP: E. Africa to Hawaii and F. Polynesia, 129 mm. White shell with yellow periostracum and axial streaks of dashed spiral lines. Animal white with red-brown streaks and blotches. Preys on fishes.

Striate Cone *Conus striatus* (cont'd) Juveniles with paler shells, animal pink-orange with black or dark brown dots. ©JJ.

Black-Band Cone *Conus sugillatus* CP: Philippines to N. Caledonia, 30 mm. Close to *C. muriculatus*, often treated as a variation. White shell with a dark area around siphon canal. Animal red with b&w speckles. ©JJ.

Thin-Line Cone *Conus tenuistriatus* IP: E. Africa to Philippines, Marshalls and F. Polynesia, 68 mm. Ovate shell, white with brown periostracum and finely granulose spiral bands. Animal red. Preys on polychaetes. ©JJ.

Tulip Cone *Conus tulipa* IP: E. Africa to F. Polynesia, 95 mm. Yellowish shell with red-brown areas and spiral rows of brown and white dots and dashes. Animal white with brown streaks and blotches. Preys on fishes. ©JJ.

Tessellate Cone *Conus tessulatus* IP: Red Sea, E. Africa and F. Polynesia, 82 mm. White shell with spiral bands of orange rectangular spots. Animal white with brown speckles and black spots. Photos: (left, right) SJ, ©JJ.

Textile Cone *Conus textile* IP: Red Sea, E. Africa to Hawaii, 150 mm. White shell with a network of brown lines, transforming into numerous tents of different size. Animal white to cream with brown mottling and black spots.

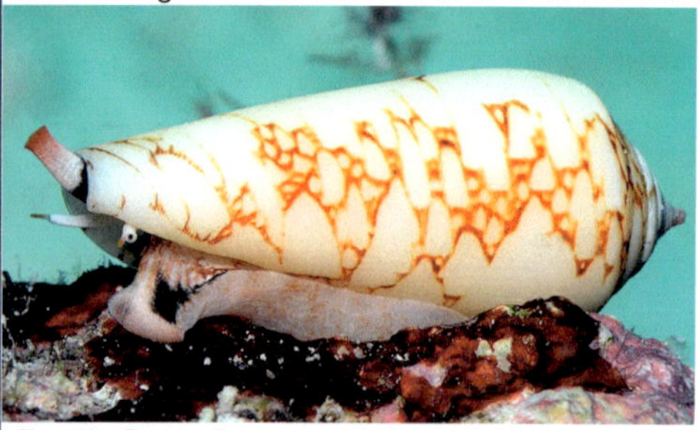

Textile Cone *Conus textile* (cont'd) Nocturnal, preys on gastropods. Photo on the left (©JJ): juvenile. Photo on the right: *Conus textile neovicarius* - Red Sea variation, to 100 mm.

Flag Cone *Conus vexillum* IP: Red Sea, E. Africa to Hawaii and F. Polynesia, 186 mm. Brown shell with white zigzag band at centre and white aperture. Foot green to greenish black. Periostracum yellowish to dark brown.

Flag Cone *Conus vexillum* (cont'd) Nocturnal, preys on eunicid worms.

Terebra Cone *Conus terebra* IP: E. Africa to Australia and Marquesas, 99 mm. Preys on terebellids. ©JJ.

Spotted Cone *Conus varius* IP: E. Africa to Marshalls, 61 mm. White shell with brown spots and spiral bands of brown and white granules. Foot white with black-brown spots. Preys on polychaetes. ©JJ.

Violet Cone *Conus viola* WP: Thailand to Philippines, Australia and Guam, 54 mm. Long narrow shell, grey to violet or reddish, with 2-3 spiral bands. Aperture violet, creamy foot. Right: © P. Bourjon, left: © Tsu Soo Tan.

Virgin Cone *Conus virgo* IP: E. Africa to F. Polynesia and Marshalls, 151 mm. Yellow to orange shell with dark-blue base, brown periostracum. Preys on terebellids. Foot white with yellow margin and black spot. Photos: ©SJ, ©JJ.

Veal Cone *Conus vitulinus* IP: Red Sea, Madagascar to Philippines and Marshalls, 83 mm. White to yellowish shel with 2 broad brown bands. Animal yellow with dark-brown spots. Preys on polychaetes. ©JJ.

Zapatos Cone *Conus zapatosensis* Philippines, 49 mm. White to cream shell with dark-brown spots, white or pale yellow aperture. White foot with brown pigment.

Spear Auger *Hastula acumen* WCP: Philippines, Indonesia to Hawaii, 36 mm. Photo on the left (©SJ): color form from Hawaii. Photo on the right: usual coloration.

White-Banded Auger *Hastula albula* IP: Red Sea, E. Africa to E. Pacific, 70 mm. ©JJ.

Red-Spot Auger *Hastula celidonota* IP: Andamans to Hawaii, 26 mm. Dorsal side with red blotch. ©JJ.

Sandbeach Auger *Hastula hectica* IP, 90 mm. Feeds on worms *Nerinides* sp.

Lance Auger *Hastula lanceata* IP: E. Africa to Hawaii and Marshalls, 76 mm. ©JJ.

Lanteri Auger *Hastula lanterii* WP: Philippines, 26 mm. Recently described by Y. Terryn from Negros, 1st time featured in situ. Distinguished from similar *H. strigilata* and *H. acumen* by the presence of 2nd band at midwhorl.

Pencil-Lined Auger *Hastula penicillata* IP: Seychelles, Madagascar to Hawaii, 40 mm. ©JJ.

Radish Auger *Hastula raphanula* IWP: E. Africa to N. Guinea and Fiji, 80 mm.

Solid Auger *Hastula solida* IP: E. Africa to Marshalls and Fiji, 35 mm. Like other family members, feeds on worms, finding them by touch of the sensitive front edge of the foot and the snout. Foot fleshy-red. ©JJ.

Strigate Auger *Hastula strigilata* IP: E. Africa to Hawaii and F. Polynesia, 56 mm. Photo on the right (©SJ): color form from Hawaii. Grey to cream midwhorl with one subsutural band.

Matheroniana Auger *Hastula matheroniana* IP: Reunion to Hawaii, 44 mm. ©SJ.

Kok's Auger *Maculauger kokiy* IWP: S. Africa to the Philippines, 86 mm. Formely *H. pseudopertusa*.

Related Auger *Myurella affinis* IP: Red Sea, East Africa to Hawaii, also E. Pacific. Like other family members, prey on worms (polychaetes) using venom to paralyze prey before it is swallowed. Photo on the right: ©JJ.

Comfortable Auger *Myurella amoena* IP: Red Sea to New Caledonia, Philippines and Hawaii, 56 mm. Animal foot pinkish dorsally, with opaque white reticulation. Right: © David Massemin.

Flour Auger *Myurella conspersa* WCP: Philippines, New Guinea to Marshalls, 53 mm. ©JJ.

Yellow-Banded Auger *Myurella flavofasciata* IP: Red Sea, E. Africa to Hawaii, 35 mm. ©SJ.

Red-Cloud Auger *Myurella nebulosa* IWP: Red Sea, E. Africa to Fiji and N. Guinea, 96 mm.

Burch's Auger *Myurella burchi* IWP: Philippines, N. Guinea to E. Africa, 40 mm. © E.G de Suduiraut.

Perforated Auger *Myurella pertusa* IP: E. Africa to N. Guinea, Marshalls and Hawaii, 115 mm. White shell with pale yellow patches, subsutural bands with pattern of dark-brown and white rectangles. Right photo ©JJ

Pygmy Auger *Myurella pygmaea* IP: Reunion to Hawaii, 13 mm. Elongated oval in shape, white to pinkish shell with contrasting dark brown spiral line just below the suture. Photo on the right - juvenile, 8 mm. ©JJ.

Column Auger *Myurellopsis columellaris* IP, 53 mm. Yellow-orange shell with white subsutural band, undulating ribs, with deep Interstices between them. Animal white. ©JJ.

Kilburn's Auger *Myurellopsis kilburni* IP: Red Sea, E. Africa to French Polynesia & Hawaii, 45 mm. White, fawn or brownish shell. Animal translucent with opaque white pigment. ©JJ.

Nathalie's Auger *Myurellopsis* cf. *nathaliae* 37 mm. WP: Philippines, similar to *M. nathaliae*, IO species.

Parkinson's Auger *Myurellopsis parkinsoni* IP: Red Sea, E. Africa to Fiji, 52 mm.

Undulate Auger *Myurellopsis undulata* IP: Madagascar to Fiji and Marshalls, 60 mm. White to yellowish shell with white subsutural band, undulating ribs with deep Interstices between them. Animal white. ©JJ.

Striped Auger *Myurellopsis paucistriata* WP: the Philippines to N. Guinea and Fiji, 40 mm. ID tentative.

Cerithlike Auger *Oxymeris cerithina* IP: Red Sea, E. Africa to Marshalls and Hawaii, 70 mm. ©JJ.

Dark-Spotted Auger *Oxymeris areolata* IP: Red Sea, E. Africa to Hawaii, 182 mm. White or pale yellow shell with 4 spiral bands of brown spots on the body whorl, and 3 on the others.

Orange Auger *Oxymeris dimidiata* IP: Red Sea, E. Africa to Marshalls and Hawaii, 166 mm. White shell with orange to red-brown maculation. Photo on the left: feeding on worm. Photo on the right ©JJ.

Tiger Auger *Oxymeris felina* IP: E. Africa to Marshalls and Hawaii, 90 mm. White shell with spiral bands of orange spots, animal white. Photos (left, right) ©JJ, © Arthur Anker.

Spotted Auger *Oxymeris maculata* IP: Red Sea, E. Africa to Hawaii and Marshalls, also E. Pacific, 275 mm. Large shell, inner lip angular in the middle. Animal white.

Crenulate Auger *Oxymeris crenulata* IP: Red Sea, E. Africa to Marshalls and Hawaii, also E. Pacific,154 mm. White shell with prominent crenulations on subsutural portion of whorl. ©JJ.

Laconic Auger *Punctoterebra succincta* IWP: Red Sea, East Africa, to Philippines, 58 mm. Animal red.

Swainson's Auger *Punctoterebra swainsoni* E. Africa to Hawaii and Tahiti, 54 mm. Shell white or brown. ©SJ.

Old Woman Auger *Terebra anilis* IWP: E. Africa to Bali, 111 mm. White shell with dark periostracum ©SJ.

Argus Auger *Terebra argus* IP: Red Sea, E. Africa to Marshalls and Hawaii, 113 mm. ©JJ.

Babylonian Auger *Terebra babylonia* IP: Red Sea, E. Africa to Hawaii and Marshalls, 83 mm. White shell, orange interspaces between the spiral grooves and axial ribs. Photo on the right: ©JJ.

Balabacen Auger *Terebra balabacensis* WP: Philippines, 25 mm. Shell grey or grey-brown.

Ringed Auger *Terebra cingulifera* IP: East Africa to Marshalls and Tuamotus, 101 mm. ©JJ.

Corded Auger *Terebra funiculata* IP: E. Africa to the Philippines, Marshalls, Hawaii and Fiji, 70 mm. White to cream or olive-brown shell, animal white. Photo on the right ©JJ.

Ringed Auger *Terebra punctatostriata* IWP: E. Africa to Philippines, 102 mm. © E.G de Suduiraut.

Montgomery's Auger *Terebra montgomeryi* IWP: N.Guinea to Marianas, 67 mm. © E.G de Suduiraut.

Jenning's Auger *Terebra jenningsi* IWP: Madagascar to Philippines and Fiji, 80 mm.

Necklace Auger *Terebra quoygaimardi* IP: E. Africa to Japan and Marshalls, 66 mm. ©JJ.

Chocolate Spotted Auger *Terebra subulata* IWP: Red Sea, E. Africa to Philippines and Fiji, 200 mm. White shell with 3 spiral bands of brown spots on the body whorl, and 2 on the others.

Amber Auger *Terebra succinea* WCP: Japan to Australia and Marshalls, 118 mm. ©SJ.

Triseriate Auger *Terebra triseriata* IWP: Reunion to India, Philippines and Hawaii, 136 mm.

Java Turrid *Turricula javana* IWP: E. Africa to India, Singapore and Australia, 70 mm. Fawn or pinkish high-spire shell with long anterior canal and spiral ridge of knobs. © Ria Tan.

White Gem-Turris *Gemmula albina* IP: Reunion to Marshalls, 75 mm. Elongated shell, white with brown dots and row of brown markings, each in the form of 3 horizontal lines, one per each sinus cord. ©JJ.

Cosmoi Gem-Turris *Gemmula cosmoi* IP: E. Africa to Japan, Philippines and Marshalls, 60 mm. Like other family members, outer lip of aperture with a notch-like sinus. ©JJ.

Necklace Turrid *Lotyrris cingulifera* IP: Red Sea, E. Africa to Vanuatu and Marshalls, 75 mm. Elongated highspire shell with relatively short siphon canal, white with with small red-brown dashes. ©JJ.

King's Turrid *Lotyrris kingae* IP: E. Africa to Philippines, Marshalls and Hawaii, 35 mm. Creamy white shell with a spiral bands of large rectangular brown maculations. ©JJ.

Short Turrid *Lophiotoma picturata* WCP: New Guinea, Philippines to Marshalls, 41 mm. Similar to *L. bratasusa*. Background color cream with distinct dark-brown blotches. Anal sinus u-shaped, deep and rather wide. ©JJ.

Marbled Turrid *Lophiotoma acuta* IP: Red Sea, East Africa to Japan and Marshalls, 90 mm. Elongated fusiform shell. Creamy-white with scattered dark-brown dots. Anterior canal moderately long. ©JJ.

White-Striped Turrid *Turridrupa albofasciata* IP: Japan to N. Caledonia and Hawaii, 22 mm. ©JJ.

Laconic Turrid *Turridrupa astricta* WCP: Philippines to Marshalls and F. Polynesia, 15 mm. ©JJ.

Bijubata Turrid *Turridrupa bijubata* IP: E. Africa to Japan, Australia and Marshalls, 25 mm. Brownish black shell with strong spiral cords, yellowish in color. Animal grey with black speckles. ©JJ.

Garnons' Turrid *Turris garnonsii* IP: E. Africa to Philippines, 75 mm. Ivory white shell with pattern of brown maculations on the subsutural fold and below the aperture. Animal white.

Babylon Turrid *Turris babylonia* WP: Philippines to Indonesia, Australia and New Caledonia, 100 mm.

Crested Turrid *Turris cristata* IWP: India to Japan, Australia and N. Caledonia, 83 mm.

Indian Turrid *Unedogemmula indica* IWP: E. Africa to Australia and Fiji, 90 mm.

Speckled Turrid *Xenuroturris millepunctata* WP: Japan, Taiwan to New Caledonia and Fiji, 50 mm.

Thick Lip Etrema *Etrema crassilabrum* WCP: Philippines to Australia, Fiji and Marshalls, 15 mm. Usually white shell with rounded whorls, operculum always absent, as for the rest of the family members. ©JJ.

Brown Etrema *Etrema* sp. CP, Marshall Islands, 6 mm. Brown shell with distinct yellow ridges. Animal with orange and yellow speckles. Similar to *E. triginostoma* (below) ©JJ. ID tentative.

Ridged Etrema *Etrema trigonostoma* WCP: Philippines to Australia and Marshall Islands, 8 mm.

Panther Glyphostoma *Glyphostoma parthenicum* SP, 8 mm. MNHN Koumac 2018. © David Massemin.

March's Lienardia *Lienardia marchei* PO, 8 mm. Guam to Loyalty Isl. and F. Polynesia, 7 mm. © Bob Abela.

Mighel's Lienardia *Lienardia mighelsi* IP: Red Sea, E. Africa, Philippines to Australia, Tahiti, 10 mm. ©JJ.

Pinkish Lienardia *Lienardia roseotincta* IP: E. Africa, Philippines to Australia and Marshalls, 20 mm. Cream to fawn shell with a tint of pink. Animal pink with white spots. ©JJ.

Red Lienardia *Lienardia rubida* IP: Red Sea to Marshalls, 13 mm. Dark bands between whorls. ©JJ.

Gilbert's Lienardia *Lienardia giliberti* WCP: the Philippines to Australia and Marshalls, 12 mm. ©JJ.

Hawk Lienardia *Lienardia vultuosa* WCP, 10 mm. White to yellow shell with prominent spiral cords. Foot white with pink network. ID tentative. ©JJ.

Dark-Banded Lienardia *Lienardia* sp. CP, Marshall Islands, 4 mm. Red shell with prominent axial ribs and spiral bands. Animal red-brown. ID tentative. ©JJ.

Yellow Lienardia *Lienardia* sp. CP, Marshall Islands, 9 mm. Visually close to to *Lienardia roseotincta* (above) but differences in shell and animal color are substantial. ©JJ. ID tentative.

Pink-Mouth Lienardia *Lienardia* sp. CP, Marshall Islands, 7 mm. ID tentative. ©JJ.

Glass-Foot Lienardia *Lienardia* sp. CP, Marshall Islands, 7 mm. Animal translucent. ID tentative. ©JJ.

Small Lienardia *Lienardia* sp. Marshall Islands, 5 mm. Animal translucent with white dots. ID tentative. ©JJ.

Banded Lienardia *Lienardia* sp. CP, Marshall islands, 7 mm. Close to *Thetidos morsura*. ID tentative. ©JJ.

Two-Bar Turrid *Clavus* sp. "*bilinearis*" CP: Hawaii, Guam, Marshall Islands, 18 mm. Distinct species, possibly described as *C. bilearis*, but looks like the description was lost. Anyway, treated as undescribed. ©JJ.

Two-Lined Turrid *Clavus bilineatus* IP: Mauritius to Philippines, Marshall Islands, New Caledonia, and Tahiti, 24 mm. Whorls often marked with double, pale and brown, interrupted line. Right photo: ©JJ.

Delphine's Turrid *Clavus delphineae* WCP: New Caledonia, Loyalty Islands and Coral Sea to Marshalls, 12.5 mm. White with a broad deep reddish-brown band around middle of last whorl. ©JJ.

Ear Turrid *Clavus canalicularis* IP: India to Japan, Australia, Fiji and Marshalls, 48 mm. White or cream, often with a broad orange-brown band at mid-last whorl. ©JJ.

Exasperating Turrid *Clavus exasperatus* IP: E. Africa, Japan to N, Caledonia and Marshalls, 25 mm. Greyishwhite, sometimes with a broad, brown band on body whorl. Animal white. Photos (left, right): ©SJ, ©JJ.

Delighted Turrid *Clavus laetus* WCP: Japan to Australia, New Caledonia, Tahiti and Hawaii, 20 mm. Glossy yellowish or orange, some dull greyish-white with blotches of yellowish. Photos (left, right): © Bob Abela, ©JJ.

Lambert's Turrid *Clavus lamberti* IP: Mauritius, Japan to Australia, Tuamotus and Marshalls, 12 mm. Glossy shell with pattern of brown spiral lines on a yellowish ground. ©JJ.

Moquini Turrid *Clavus moquinianus* WCP: N. Caledonia, Marshalls, 5 mm. White with blotch of brown ©JJ.

Small Clavus *Clavus pica* WCP, 25 mm. Similar to *C. moquinianus* but has 18-21 axial ribs instead of 12 ©JJ.

Smooth-Ribbed Clavus *Clavus obliquicostatus* WCP: Indonesia, Philippines to Marshalls, 6 mm. Variable in color: white shell, tinged between ribs with brown. Right: © David Massemin, MNHN Koumac 2018.

Smooth Turrid *Clavus* sp. New Caledonia, 14 mm. White shell with inconspicuous axial ribs. ID tentative.

One-Zoned Turrid *Clavus unizonalis* E. Africa, Japan to Australia and Marshalls, 26 mm. ©JJ.

One-Zoned Turrid *Clavus unizonalis* (cont-d) Variable in color and in morphology. Typical *C. unizonalis* with rounded ribs (above), this form with angular ribs - *C. vidua* - recently was synonymised with *C. unizonalis.* ©JJ.

One-Zoned Turrid *Clavus unizonalis* (cont-d) Animal variable in color, translucent to white with black speckles, siphon black. Genetic data may help with species delimitation in this *Clavus unizonalis/C. viduus* group.

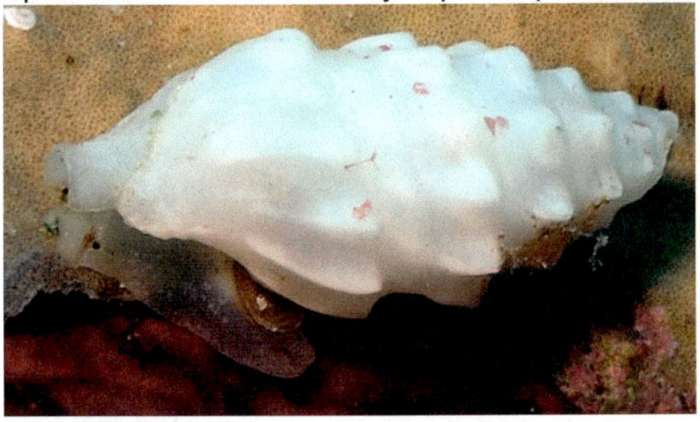

White Clavus *Clavus* sp. CP: Marshall Islands, 9 mm. Animal translucent white. ID tentative. ©JJ.

Wilmer's Tirrid *Clavus* cf. *wilmeri* CP: Marshall Islands, 22 mm. ID tentative. ©JJ.

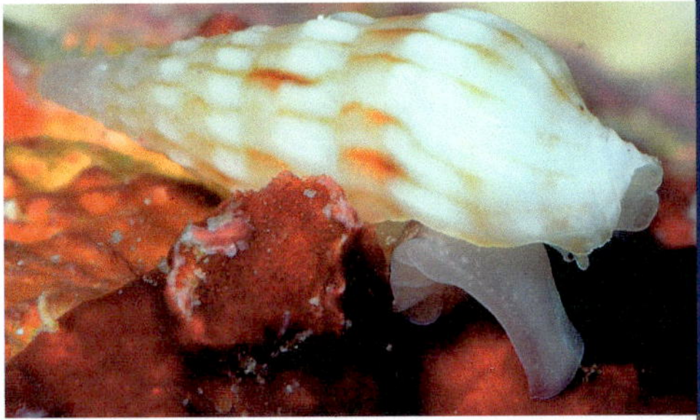

White Clavus *Clavus* sp. CP: Marshall Islands, 12 mm. Glossy shell with smooth axial ribs, white with pattern of pale brown bands and diffuse dark-brown blotch on body whorl. Animal translucent white. ID tentative. ©JJ.

DRILLIIDAE

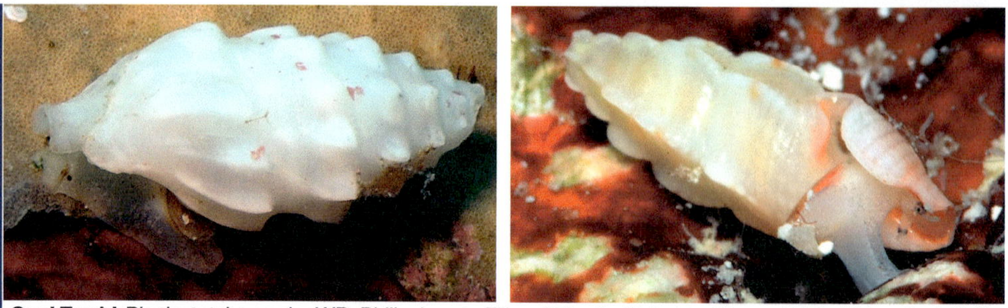

Opal Turrid *Plagiostropha opalus* WP: Philippines to Fiji and Marshall Islands, 20 mm. White high spired shell with pentagonal shape in transverse section. Animal white with grey speckles. ©JJ.

Fivefold Turrid *Plagiostropha quintuplex* WP: Philippines to Taiwan and Marshall Islands, 8 mm. Glossy high spired shell, yellow to red-brown in color, with pentagonal shape in transverse section. ©JJ.

HORAICLAVIDAE

Pinkish Carinapex *Carinapex* sp. CP: Marshall Islands, 6 mm. ID tentative. ©JJ.

Brownish Carinapex *Carinapex* sp. CP: Marshall Islands, 6 mm. ID tentative. ©JJ.

Pigmy Haedropleura *Haedropleura pygmaea* WCP: Taiwan to Australia, Fiji and Marshall Islands, 5 mm. White shell with brown bands on the body whorl. Animal translucent with black speckling. ©JJ.

Obscure Haedropleura *Haedropleura* sp. CP: Marshall Islands, 5 mm. Cream shell with pattern of brown bands, axial ribs distinctly curved. Animal translucent with dark grey speckling. ©JJ.

Queer Haedropleura *Haedropleura* sp. CP, Marshalls, 5 mm. Close to *H. pigmaea*, but fewer axial ribs. ©JJ.

Regal Turrid *Vexitomina regia* IP: Mauritius to Philippines and N. Caledonia, 50 mm. © D. Massemin

Angel Eucithara *Eucithara angela* WCP: Australia, Vanuatu to Marshall Islands, 16 mm. White turreted shell with axial ribs, crossed by the faint spiral striations. Animal white, marbled with red and grey. ©JJ.

Sand-Crawling Turrid *Eucithara* cf. *arenivaga* WP: Indonesia, 15 mm. Turreted shell with rounded axial ribs, crossed by the brown spiral bands. Animal creamy, spotted with orange.

Hairy Turrid *Eucithara capillata* WCP: Philippines to Australia, New Caledonia and Marshalls, 8 mm. White turreted shell with fusiform axial ribs. Animal white, marbled with red. ©JJ.

Fusiform Turrid *Eucithara fusiformis* IP: E. Africa, Oman, Philippines to Guam, 15 mm. © Bob Abela.

Pulchell's Turrid *Eucithara pulchella* WP: Australia to Philippines, Indonesia, Tonga, 16 mm.

Some examples of possibly undescribed *Eucithara* sp. from the Central Pacific, Marshalls. All 4-6 mm, except upper right and left, 9-10 mm. All photos ©JJ, except bottom right: © Christina Sylvester.

"The positive identification of most Indo-Pacific mangeliines and often even their allocation to modern genera - is more often than not hampered by the meagerness of the original descriptions and figures... *Eucithara* has been used as a convenient dumping ground for a large number of Indo-Pacific species." R. N. Kilburn, 1992

Conch-Like Turrid *Eucithara stromboides* IP: Reunion, Philippines to Indonesia, Australia, Marshalls, 25 mm. Biconic to fusiform shell. Axial ribs more or less strong, crossed by the fine to coarse spiral threads.

Conch-Like Turrid *Eucithara* cf. *stromboides* WCP: Indonesia, Philippines to Marshalls, 25 mm. Probably variation of *E. stromboides*, but siphon with distinct orange and white dots. ID tentative.

Cylindrical Turrid *Gingicithara cylindrica* IP, 18 mm. © Philippe & Guido Poppe www.poppe-images.com

Chinese Macteola *Macteola chinoi* WCP: Philippines, Marshalls, 10 mm. ©JJ.

Chinese Macteola *Macteola chinoi* (cont-d) Animal white with fine red-orange reticulation and marginal translucent patches.

Interrupted Macteola *Macteola interrupta* IP: Reunion, Philippines to Australia, Marshall Islands and Hawaii, 12 mm. Biconical shell, body whorl with dotted spiral striation. Animal yellowish with red dots. ©JJ. ID tentative.

Greenish Macteola *Macteola* sp. CP: Marshall Islands, 5 mm. Similar to M. chinoi, but animal colored differently, pale greenish with orange spots. ©JJ.

Banded Macteola *Macteola* sp. CP: Marshall Islands, 5 mm. Similar to M. interrupta, but animal colored differently, white with red reticulation. ©JJ.

Black Lovellona *Lovellona atramentosa* IP: E. Africa, Philippines, Hawaii, Polynesia and Marshalls, 13 mm. Inflated dark-brown or black shell, was described as Conus in 1849. White animal with red head and red siphon.

Purplish Turrid *Mitromorpha alphonsiana* WCP: New Caledonia, Marshalls, 5 mm. © S. Jazwinski.

Pale Turrid *Mitromorpha* sp. CP, Marshalls, 9 mm. Possibly smoother form of *M. flammulata* (below) ©JJ.

Twinkle Turrid *Mitromorpha flammulata* WCP: Japan, Guam, Marshalls, Marianas, 9 mm. Translucent animal with yellow dotting. ©JJ.

Golden Daphnella *Daphnella aureola* WCP: the Philippines, Hawaii, Marshalls. 22 mm. ©JJ .

Flame Daphnella *Daphnella flammea* IP: Reunion, New Guinea, Philippines and Marshalls, 15 mm. ©JJ.

Scraper Daphnella *Daphnella radula* WCP: Japan, Philippines, Marshall Islands, 14 mm.

Reeve' Daphnella *Daphnella reeveana* IP: Reunion, Philippines, Guam, 8 mm. © Bob Abela.

Rissoid-Like Daphnella *Daphnella rissoides* IP: E. Africa, Philippines to New Caledonia and Marshalls, 20 mm. Variable in color, creamy to red-brown shell. Easy recognizable by spiral bands of white dots. ©JJ.

Varicose Daphnella *Daphnella varicosa* WP: Taiwan, New Caledonia, Hawaii and Marshalls, 16 mm. ©JJ.

Rough Daphnella *Daphnella* sp. CP: Marshall Islands, 8 mm. Nocturnal, found in cave. ©JJ.

Wheat-Grain Diaugasma *Diaugasma olyra* WCP: Philippines to Taiwan and Marshalls, 16 mm. ©JJ.

Apiculate Turrid *Hemilienardia apiculata* WCP: Australia, N. Caledonia, Marshalls, 8 mm. ©JJ.

Mauve Turrid *Hemilienardia goubini* WCP: Philippines to Australia and N. Caledonia, Marshalls, Loyalty Isl, 5 mm. Inflated fusiform shell with stout axial ribs and spiral threads. Translucent animal, orange siphon. ©JJ.

Mallet Turrid *Hemilienardia malleti* IP: Red Sea, E. Africa, Philippines, Marshalls, 6 mm. Variable in color shell, pinkish to mauve and cherry with white bands. Convex whorls, stout axial ribs and spiral threads. ©JJ.

Eyed Turrid *Hemilienardia ocellata* IP: Mauritius to Maldives, Philippines to Marshalls, 11 mm. ©JJ.

Lockers Turrid *Hemilienardia thyridota* WCP: Philippines to Australia, Marshalls, 6 mm. ©JJ.

Red & White Turrid *Hemilienardia* sp. Marshall Islands, 5.5 mm. Shell with convex whorls, stout axial ribs. Red with white spiral bands, combined of 4 spiral threads. Animal pink with white dots. ©JJ.

Tiny Kermia *Kermia* cf. *benhami* CP: Marshalls, 4 mm. Resembles elongated *K. benhami.* © C. Sylvester.

Brown-Dotted Kermia *Kermia* sp. CP: Marshalls, 5 mm. Grainy shell, animal with white dots. ©JJ.

Pink Kermia *Kermia* sp. CP: Marshalls, 5 mm. Elongated grainy shell. © C. Sylvester. ID tentative.

Pink Kermia *Kermia* sp. CP: Marshalls, 10 mm. Elongated grainy shell. © C. Sylvester. ID tentative.

Barnard's Turrid *Pseudodaphnella barnardi* IP: Seychelles, Australia, Marshalls, 5 mm. Genera Kermia and Pseudodaphnella are differentiated only in protoconch morphology, molecular studies doesn't support it.

Daedal's Turrid *Pseudodaphnella daedala* WCP: Australia, Marshalls, 5 mm. © Christina Sylvester.

Red-Spotted Turrid *Pseudodaphnella rubroguttata* WCP: Australia, Marshalls, 5 mm. ©JJ.

Grain-Ribbed Turrid *Pseudodaphnella granicostata* IP: E. Africa, Philippines, Indonesia, Marshalls, Guam, 13 mm. Easily recognizable shell with red, orange and white nodules. © Christina Sylvester.

Interwoven Turrid *Pseudodaphnella nexa* WCP: Philippines to New Caledonia, Marshalls, Tahiti, 20 mm. White elongated high spired shell, distinctly reticulated with spiral cords and axial ribs. ©JJ.

Cryptic Turrid *Pseudodaphnella* cf. *crypta* Marshalls, 5 mm. Poss. immature specimen of *P. crypta.* ©JJ

Netted Turrid *Pseudodaphnella* sp. CP: Marshalls, 5 mm. © Christina Sylvester. ID Tentative.

Cloudy Turrid *Pseudodaphnella* cf. *nebulosa* CP: Marshalls, 11 mm. White elongated shell, axial ribs with prominent nodules. Morphology similar to *P. nebulosa*, species from Hawaii and F. Polynesia. ©JJ.

Speckled Turrid *Pseudodaphnella tincta* IP: Seychelles, Japan to Australia, Marshalls, 10 mm. ©SJ.

Cuming's Tritonoturris *Tritonoturris cumingii* IP: Red Sea, E. Africa, Indonesia to Hawaii, 40 mm. SJ.

Cuming's Tritonoturris *Tritonoturris cumingii* (cont'd) Easily recognizable species, yellow-brown shell with inflated globose whorls and checkered pattern of pale spiral cords and axial ribs.

Pink Tritonoturris *Tritonoturris menecharmes* IP: Mauritius, Philippines, Marshalls, 30 mm. Pale brown to pink shell with inflated whorls. Translucent animal white dots, tentacles with red bands. Photos (left, right:) ©SJ, ©JJ.

Ladder Crassispira *Crassispira scala* WCP: N. Guinea, N. Caledonia to Philippines, Marshalls, 30 mm. ©JJ.

Humpy Inquisitor *Inquisitor tuberosus* WP: Taiwan, Philippines, 18 mm.

Spotted Inquisitor *Inquisitor sterrha* WCP: Australia, Philippines, 40 mm. Axial ribs with darker interstices.

Red & White Naudedrillia *Naudedrillia* sp. CP: Marshalls, 9 mm. ID tentative. ©JJ.

Strong Mitre *Calcimitra salva* Philippines, N. Caledonia, 91 mm. Apricot to brownish with white lines.

Exceptional Mitre *Carinomitra peculiaris* WP: New Guinea to Philippines and Marshalls, 15 mm. ©JJ.

Dancing Mitre *Carinomitra saltata* CP: F. Polynesia, Hawaii, Marshalls, 6 mm. White to brownish shell. ©JJ.

Cany Mitre *Carinomitra typha* IP: Red Sea, Reunion to Japan and Marshalls, 14 mm. ©JJ.

Bernhardina Knobby Mitre *Condylomitra bernhardina* IP: E. Africa to Philippines, Australia and Marshalls, 30 mm. Variable in color, yellow to red brown shell, last 3 whorls produced into a pointed spire. Animal white. ©JJ

Horn-Colored Mitre *Domiporta carnicolor* IP: Red Sea, East Africa, Hawaii, Marshalls, 38 mm. ©JJ.

File Mitre *Domiporta filaris* IP: Red Sea, Philippines, Marshalls, Hawaii, 55 mm. ©JJ.

Glorious Mitre *Domiporta gloriola* WCP, 67 mm. © Philippe & Guido Poppe www.poppe-images.com.

Flecked Mitre *Domiporta granatina* IP: Red Sea, E. Africa, Japan to Hawaii, 74 mm.

Excellent Mitre *Domiporta praestantissima* IP: Red Sea, Madagascar to F. Polynesia, 61 mm. ©JJ.

Ringed Mitre *Imbricaria annulata* IP: Red Sea, E. Africa, Japan to Guam, 35 mm. © Bob Abela.

Armonica Mitre *Imbricaria armonica* WP: Philippines, Indonesia, 41 mm.

Flash Mitre *Imbricaria fulgetrum* IP: Reunion, Philippines to Marshalls, 28 mm. ©JJ.

Cone Mitre *Imbricaria conularis* IP: Mauritius to Japan, Australia, Marshalls and F. Polynesia, 26 mm. Conically ovate shining shell, marbled with white and grey. Left: MNHN Koumac 2018, © David Massemin, right: ©JJ.

Verrucous Mitre *Imbricaria verrucosa* WCP: N. Guinea to Philippines, Marshalls, 38 mm. Shell with prominent spiral ridges, axial ribs and sharp spinose nodules at intersections. Translucent animal, marbled with white. ©JJ.

Salisbury's Mitre *Imbricaria salisburyi* IP, 25 mm. © Philippe & Guido Poppe www.poppe-images.com.

Bonelike Mitre *Imbricariopsis punctata* IP, 25 mm. Yellowish or fawn in color, early whorls whitish. ©JJ.

Conovula Mitre *Imbricariopsis conovula* IP: Mauritius to Australia, Marshalls and Hawaii, 20 mm. Similar to *I. punctata,* differs in being less globose near the shoulder, anterior canal always dark-purple. ©JJ.

Vanikora Mitre *Imbricariopsis vanikorensis* IP: Red Sea to Japan and Marshalls, 20 mm. Variable in color: cream to brownish, scattered red dots with white flames, early whorls whitish. Animal white with brown patches. ©JJ.

Giant Mitre *Mitra mitra* IP: Red Sea, E. Africa-Japan, Hawaii and Marshalls.180 mm. Large shell, white with 5-9 transverse rows of orange spots and 5-8 punctate spiral lines. Creamy white animal, edges brown. ©JJ.

Papal Mitre *Mitra papalis* IP: E. Africa to Japan, Marshalls and Hawaii, 166 mm. Ovate shell, whitish in color, with close-set transverse rows of red-brown blotches. Animal cream. ©JJ.

Pontifical Mitre *Mitra stictica* IP: E. Africa to Fiji, F. Polynesia and Hawaii, 87 mm. Large shell, white or cream in color, with rows of close-set irregular orange blotches. ©JJ

Swollen Mitre *Mitra turgida* IP: IWP: Madagascar to Philippines and Guam, 16 mm. © Bob Abela.

Truncate Mitre *Nebularia coarctata* IP: Reunion, Philippines to Australia and Marshalls, 34 mm. ©JJ.

Acuminate Mitre *Nebularia acuminata* IP: E. Africa to Japan, F. Polynesia and Hawaii, 35 mm. Elongate-ovate, cream or yellow shell with acuminate spire. Animal yellowish. ©JJ.

Gold-Mouth Mitre *Nebularia chrysostoma* IP, 66 mm. White to yellow shell with irregular dark blotches, body whorl often with pale central transverse zone. Cream animal with brown blotches. ©JJ.

Toothless Mitre *Nebularia edentula* IP: Reunion to Japan and Marshalls, 40 mm. ©SJ.

Top Mitre *Nebularia fastigium* IP: E. Africa to Philippines, Japan and Marshalls, 21 mm. ©SJ.

Adusta Mitre *Nebularia eremitarum* IP: 86 mm. © Philippe & Guido Poppe www.poppe-images.com.

Pyramid Mitre *Nebularia pyramis* IP: Reunion to Guam and Marshalls, 50 mm. © Bob Abela.

Rusty Mitre *Nebularia ferruginea* IP: E. Africa to Fiji, Hawaii, F. Polynesia, 58 mm. Whitish to cream shell with elevated spiral ridges. Decorated with rusty-brown blotches, often wit a pale central zone on the body whorl. ©JJ.

White Mitre *Nebularia nivea* WCP: Japan to Philippines, Tuamotu and Hawaii, 90 mm. Photo on the right: feeding on sipunculid worm, usual diet for all family members. ©SJ.

Scorched Mitre *Nebularia ustulata* IP: Reunion to India, Philippines and Marshalls, 62 mm. Elongated shell, cream to yellow in color, with irregular brown blotches and fine spiral striae. Animal white. ©JJ.

Snake-Tongue Mitre *Nebularia pellisserpentis* IP: Mauritius, Japan, Philippines to Hawaii, 35 mm. ©SJ.

Chrysalis Mitre *Pseudonebularia chrysalis* IP: Red Sea, E. Africa to Japan and Marshalls, 30 mm. ©JJ.

Armadillo Mitre *Pseudonebularia* cf. *cingulata* WCP: Marshalls, 50 mm. Orange to brown shell with elevated spiral ridges. Animal translucent with opaque yellow spots. Appears to be undescribed. ©JJ.

Kettle Mitre *Pseudonebularia cucumerina* IP: Red Sea, E. Africa to Hawaii, Polynesia, 26 mm. Yellowish to red-brown in color, often with a white transverse band on the body whorl. Animal with yellow spots. ©JJ.

Barrel Mitre *Pseudonebularia doliolum* IWP: India to Japan, Australia and New Guinea, Marshalls, 30 mm. Yellow, brown or red-brown shell with elevated spiral ridges. ©JJ.

Barrel Mitre *Pseudonebularia doliolum* (cont-d) Redbrown color morph.

Strawberry Mitre *Pseudonebularia fraga* IP: Red Sea, to Solomons, N. Guinea and Marshalls, 50 mm. ©JJ.

Kantor's Mitre *Pseudonebularia kantori* WCP: Philippines; Marshall Islands, 17 mm. ©JJ.

Lienard's Mitre *Pseudonebularia lienardi* IP: E. Africa to Marshall Islands and Hawaii, 32 mm. ©JJ.

Rubiginea Mitre *Pseudonebularia rubiginea* WCP: Australia to Philippines and Hawaii, 45 mm. Elongate-ovate red shell, with wide-spaced spiral ridges and rounded axial mounds. White animal with yellow spots. ©JJ.

Red-Painted Mitre *Pseudonebularia rubritincta* IP: Reunion to Japan, Philippines, Fiji and Marshalls, 35 mm. Red shell with white transverse bands, widest on the the body whorl. Yellow animal with paler dots. ©JJ.

Sarino's Mitre *Pseudonebularia sarinoae* Philppines to Marshalls, 11 mm. Paler and darker bands. ©JJ.

Silvia's Mitre *Pseudonebularia silviae* IP: Maldives to Malaysia, Indonesia, Marshalls, 13 mm. ©JJ.

Clathrus Mitre *Neocancilla clathrus* IP: Red Sea, E. Africa to Japan, Fiji and Marshalls, 64 mm. Variable in color, white or creamy with 1-2 irregular red-brown transverse bands on body whorl. ©JJ.

Butterfly Mitre *Neocancilla papilio* IP: Red Sea, E. Africa to Marshalls, 64 mm. Whitish to pale cream in color, with 1-2 broad, reddish transverse bands. Discharge, when disturbed, a purple protective mucus. ©JJ.

Crenulate Mitre *Pterygia crenulata* IP: Red Sea, E. Africa to Japan and Hawaii, 41 mm. ©JJ.

Fenestrate Mitre *Pterygia fenestrata* IP: Red Sea to Australia and Tuamotu, 40 mm. © D. Massemin.

Finger Mitre *Pterygia dactylus* IP: India to Japan, Australia, Marshalls, 55 mm. Whitish to cream in color with 3-4 yellow or brown transverse bands on the body whorl. Animal white. Photos (left, right) ©SJ, ©JJ.

Nut Mitre *Pterygia nucea* IP: Red Sea, E. Africa to Fiji, Polynesia and Marshalls, 72 mm. White or cream shell with 1-2 brown transverse bands and spiral rows of small brown spots. Animal white.

Cardinal Mitre *Quasimitra cardinalis* IP: E. Africa, to Marshalls and F. Polynesia, 90 mm. ©JJ.

Particolored Mitre *Quasimitra nubila* IP: Red Sea, E. Africa to Fiji and Marshalls, 81 mm. ©SJ.

Diadem Mitre *Quasimitra puncticulata* IP, 57 mm. © Philippe & Guido Poppe www.poppe-images.com.

Bloodstained Mitre *Quasimitra sanguinolenta* IO: Red Sea, E. Africa, 42 mm.

Speckled Mitre *Roseomitra millepunctata* IWP, 24 mm. © Philippe & Guido Poppe www.poppe-images.com.

Rosacea Mitre *Roseomitra rosacea* IWP: S. Africa to Japan and N. Caledonia, 48 mm. © D. Massemin.

Desetang's Mitre *Scabricola desetangsii* IP: Red Sea and S. Africa to Samoa, Guam and Marshalls, 35 mm. White to cream shell with elevated spiral ridges and irregular darker bands. Photos (left, right) © Bob Abela, ©JJ.

Olive-Shaped Mitre *Scabricola olivaeformis* IP: Reunion to Japan, Hawaii and Polynesia, 21 mm. Shell: yellow in color, with a dark purple apex and anterior canal. Animal white with brown tentacles.

Variegated Mitre *Scabricola variegata* IP: Red Sea, E. Africa to Marshalls and Polynesia, 50 mm. Shell: white in color, with irregular longitudinal brown blotches. Animal white with yellow and brown mottling. ©JJ.

Victor Dan's Mitre *Scabricola vicdani* WP: Philippines to New Guinea and Solomons, 50 mm. ©SJ.

Imperial Mitre *Strigatella imperialis* IP: Red Sea to Fiji, Solomons, Marshalls, 65 mm. ©JJ.

Dawn Mitre *Strigatella aurora* IP: Red Sea to Marshalls and Hawaii, 50 mm. Shell red or brown in color, with large irregular white blotches. Animal red-brown with white edges. ©JJ.

Coffee Mitre *Strigatella coffea* IP: E. Africa to Japan, Polynesia and Hawaii, 60 mm. Coffee-brown or red-brown shell, animal red-brown, tentacles with white tips. Photos (left, right): © Philippe Bourjon, ©JJ.

Shortened Mitre *Strigatella decurtata* WCP: Japan to Marshalls and Polynesia, 37 mm. Ovate heavy shell, dark brown with irregular white spots. Animal brown. ©JJ.

Tawny Mitre *Strigatella fulvescens* IP: Reunion to Japan, Philippines, Marshalls and Tuamotu, 50 mm. Variable in color: yellow to red-brown or brown shell with longitudinal rows of yellow dots. Animal red-brown. ©JJ.

Lettered Mitre *Strigatella litterata* IP: Red Sea to Fiji, Polynesia and Hawaii, 35 mm. ©JJ.

Poverty Mitre *Strigatella paupercula* IP: E. Africa to Fiji, Samoa and Marshalls, 40 mm. ©JJ.

Mournful Mitre *Strigatella luctuosa* IP: E. Africa to Japan, Australia, Marshalls, 35 mm. ©JJ.

Magpie Mitre *Strigatella pica* IP: Mauritius to Solomons, 45 mm. Coffee-brown elongate shell.

Affected Mitre *Strigatella vultuosa* WCP: Philippines, New Caledonia, Hawaii, 20 mm. ©SJ.

Polished Swainsonia *Swainsonia limata* IP: Mauritius to Philippines and Marshalls, 30 mm. ©JJ.

Chaste Mitre *Swainsonia casta* IP: East Africa to Japan, Indonesia, Marshalls and Polynesia, 55 mm. Elongate shell with pointed spire, White with a broad chestnut-brown band on the body whorl. ©JJ.

Newcomb' Mitre *Swainsonia newcombii* WCP, 45 mm. © Philippe & Guido Poppe www.poppe-images.com.

Eyed Mitre *Swainsonia ocellata* IP: E. Africa to Japan, Philippines and Australia, 34 mm. ©SJ.

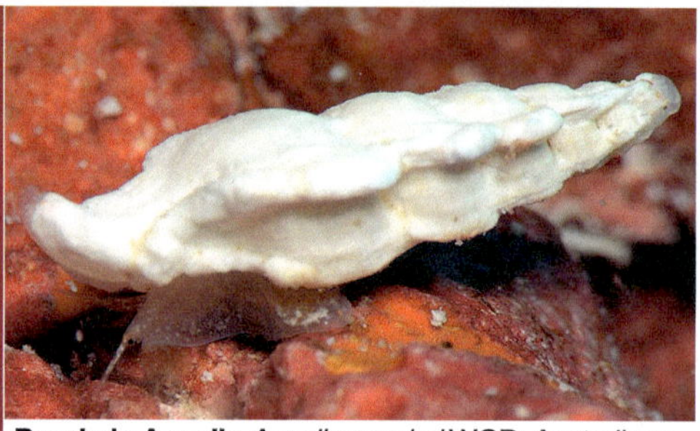

Ponder's Aspella *Aspella ponderi* WCP: Australia to Japan and Marshall Islands, 18 mm. ©JJ.

Elongate Aspella *Aspella producta* IP: Red Sea, E. Africa to Marshalls and Hawaii, 33 mm. ©JJ.

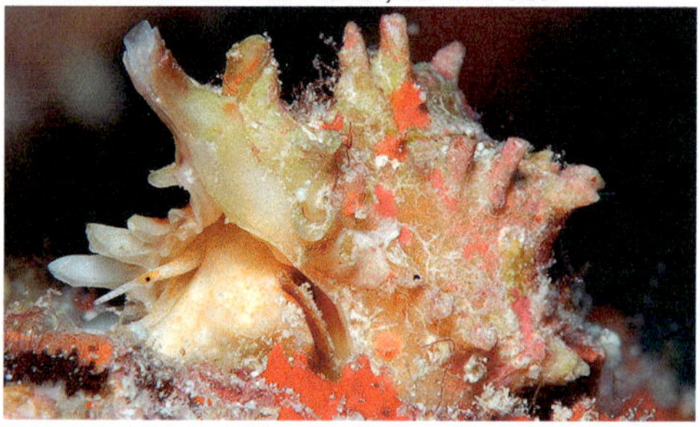

Caledonia Attiliosa *Attiliosa caledonica* WCP: New Caledonia to Marshall Islands, 31 mm. Shell with 4 or 5 distinct acute spines on the varices of the body whorl (highlighted). Animal cream with orange speckles. ©JJ.

High Murex *Chicomurex excelsus* WCP: Philippines to Fiji and Marshall Islands, 51 mm. ©SJ.

False Superb Murex *Chicomurex pseudosuperbus* IP, 60 mm. MNHN Koumac 2019. © D. Massemin.

Laciniate Murex *Chicomurex laciniatus* WCP: Philippines to Australia and Marshalls, 77 mm. High spired shell, finely dentate outer lip, fleshy ochre or brown in color. Aperture white with deep rosy pink to violet edges. ©JJ.

Tagaroa Murex *Chicomurex tagaroae* WP: Philippines, 50 mm. High spired shell, nearly round aperture, dentate outer lip, tan or brown in color.

Banks' Murex *Chicoreus banksii* IWP, 107 mm. MNHN Koumac 2018. © D. Massemin.

Insular Murex *Chicoreus insularum* Hawaii, 100 mm. Light brown, dark brown on the spiral elements. ©SJ.

Burnt Murex *Chicoreus brunneus* IP: E. Africa to Indonesia, 115 mm. Brown to orange color, with dark transverse bands. Shoulder spines elongate and foliate; canal with three spines. ©JJ.

Ingrid-Maria Murex *Chicoreus ingridmariae* WCP: Indonesia to Marshalls and Tahiti, Bluish aperture, violet lips. Broadening canal with three spines. Yellow or orange in color with violet transverse bands. ©SJ.

Jessica's Murex *Chicoreus jessicae* WP: Philippines, Indonesia, 80 mm. Tan or brown with frondose spines.

Tom Pain' Murex *Chicoreus paini* WP: Indonesia to Solomons and New Caledonia, 80 mm.

Fluted Murex *Chicoreus laqueatus* WCP: Japan to Hawaii, 40 mm. Easy to recognize small species, inner surface of the outer lip with 11 denticles. Shell color is pinkish, with areas of deeper rose-pink. ©SJ.

Rose-Branch Murex *Chicoreus palmarosae* IP: E. Africa to Japan, 130 mm. Large, easy to recognize species: body whorl with 4 broad, foliose spines. A wide gap divides the longest, shoulder spine and 3 shorter spines.

Rose-Branch Murex *Chicoreus palmarosae* (cont-d) Canal with 3 spines. Shell color is red to brown, spiral elements darker. Feeds on bivalves and other gastropods. ©JJ.

Ramose Murex *Chicoreus ramosus* IWP: Red Sea to Japan, N. Caledonia and Polynesia, 342 mm. Large shell, white in color. Leaf-like, moderately short spines, longest and strongest at shoulder.

Saul's Murex *Chicoreus saulii* IP: Reunion to Japan, New Guinea and Marshalls, 142 mm. Shell color is orange-brown. Spines arc with 4 primary and 5 secondary spines, as shown. Animal red with white speckles. ©JJ.

Firebrand Murex *Chicoreus torrefactus* IP: E. Africa, to India, Marshalls, Tahiti. 160 mm. Outer lip with a large shoulder spine, followed by 4 shorter spines, secondary spines (spinelets) between them. Animal brownish. ©JJ.

Virgin Murex *Chicoreus virgineus* IO: Red Sea to India, 160 mm. White aperture, rimmed with pink.

Carole's Coralliophila *Coralliophila caroleae* IP: East Africa, Japan to Philippines, Marshalls. © J. Lord.

Amirantes Coralliophila *Coralliophila amirantium* IP: Seychelles to Philippines and Marshall Islands. Found on Seriatopora corals, imitates sponge by shell structure and color. Animal orange with white spots. ©JJ.

Ribbed Coralliophila *Coralliophila costularis* IWP: Red Sea, Maldives, Australia, 60 mm. Interior violet to purple.

South Seas Coral Shell *Coralliophila erosa* IP: Red Sea to Hawaii, 35 mm. On *Porites* corals. ©JJ.

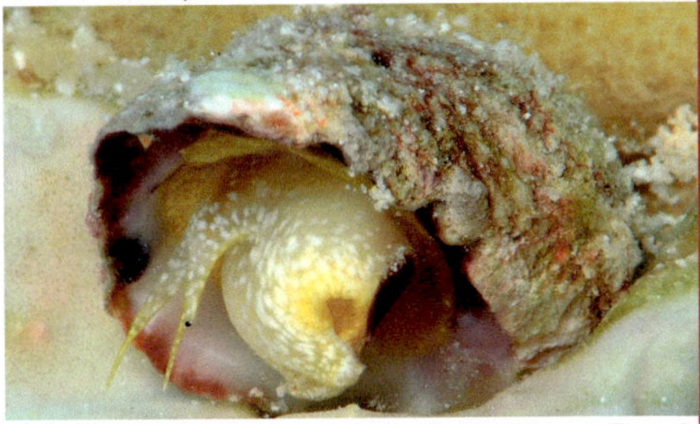

Quoy's Coral Shell *Coralliophila monodonta* IP: E. Africa, Maldives to N. Guinea and Marshalls, 35 mm. Found on *Madrepora* corals. Globose shell, white aperture with dark purple anterior margin. Animal yellow. ©JJ.

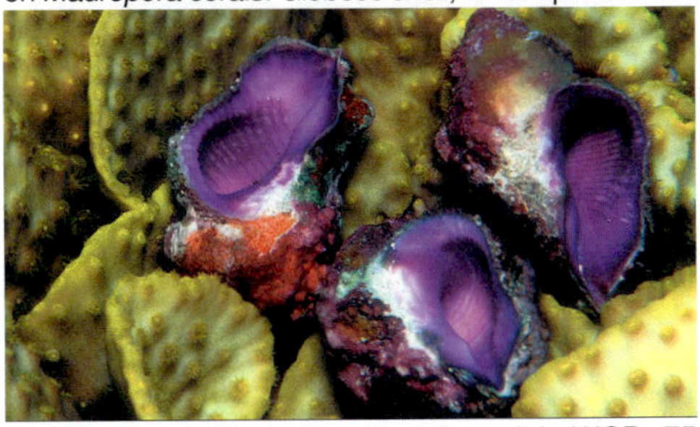

Radula Coral Shell *Coralliophila radula* WCP, 75 mm. White, interior violet, on *Turbinaria* coral. ©JJ

Grid Coral Shell *Coralliophila clathrata* IP: E. Africa to Polynesia, 15 mm. *Palythoa* zoanthid symbiont.

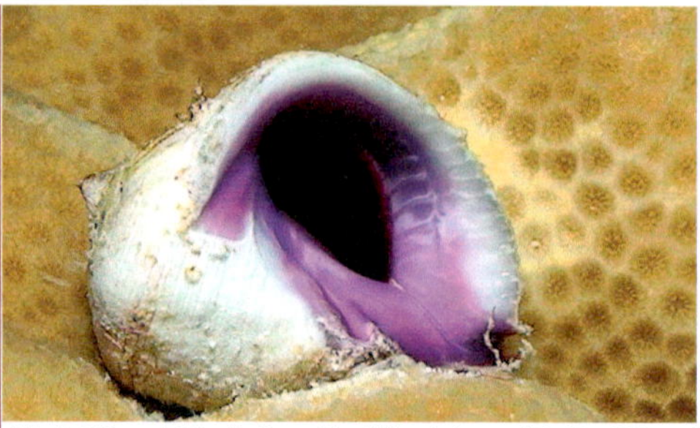

Violet Coral Shell *Coralliophila violacea* IP: Red Sea, E. Africa to Marshalls, also Galapagos, 47 mm. Globose shell, white to grey in color, interior violet. Found on massive corals like Porites. Photos (left, right): ©SJ, ©JJ.

Lefevre' Morula *Cytharomorula lefevreiana* IP: Red Sea to Japan and Marshall Islands, 14 mm. Small shell, white in color, shouldered whorls, obvious spiral cords. Translucent animal with yellowish spotting. ©JJ.

Translucent Morula *Cytharomorula* sp. CP: Marshall Islands, 5 mm. © Christina Sylvester.

Dentate Drupe *Drupa denticulata* IO: Madagascar, Reunion, 47 mm. © Philippe Bourjon.

Spotted Drupe *Drupa ricinus* IP: Mauritius to Maldives, Philippines to Marshalls and Guam, 32 mm. Outer lip inside with two sets of teeth arranged in four and two (highlighted). Photos (left, right) ©JJ, © Bob Abela.

Raspberry Drupe *Drupa rubusidaeus* IP: Red Sea, E. Africa to Hawaii and Marshalls, 60 mm. Shell with 5 rows of spines, Interior of the aperture light purple. Animal greenish. Photos (left, right) ©SJ, ©JJ.

Horn Drupe *Drupella cornus* IP, 40 mm. E. Africa to Hawaii. Feeds on *Acropora* and *Montipora* corals. ©JJ.

Strawberry Drupe *Drupella fragum* IP: East Africa to Hawaii, 30 mm. Feeds on branching corals. ©SJ.

Shouldered Castor Bean *Drupella margariticola* IP: Red Sea, E. Africa to Marshalls, 40 mm. ©JJ.

Rugose Drupe *Drupella rugosa* IP: Marshalls to Red Sea, 35 mm. Body whorl with 5 rows of tubercles. ©SJ.

Finger Drupe *Drupina grossularia* IP, 45 mm. Easily recognizable species due to golden yellow aperture and outer lip processes, starting with 2 longest, similar to V sign. Animal yellow with small white spots. ©JJ.

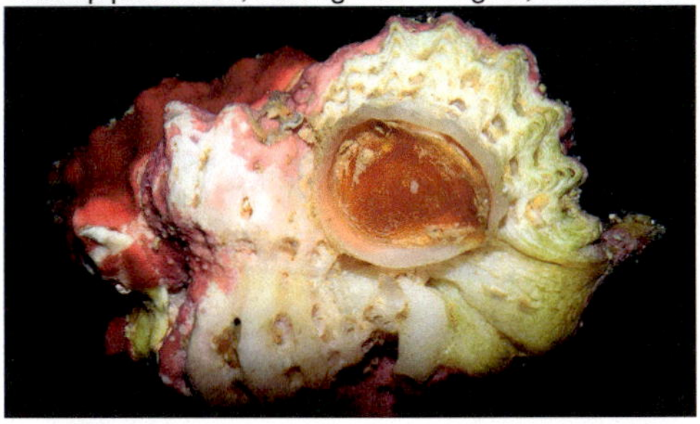

Short Favartia *Favartia brevicula* IWP: E. Africa to Australia, Fiji and N. Caledonia, 35 mm. Small shell, perfect coral rubble mimic. Pale purplish aperture, upturned siphon canal. Left photo MNHN Koumac 2018, right ©JJ.

Crouch' Favartia *Favartia crouchi* IP: Mauritius to Marshalls and Tuamotu, 12 mm. ©JJ.

Round-Mouth Fawartia *Favartia cyclostoma* IP: Red Sea, E. Africa, Philippines to Marshalls, 11 mm. ©JJ.

Guam Favartia *Favartia guamensis* IP: E. Africa to Polynesia and Marshalls, 12 mm. Yellow or orange in color, elongated small shell. Five strong spiral cords, squarish in cross-section. Animal white. ©JJ.

Banded Murex *Muricidae* sp. CP: Marshalls, 13 mm. Genus unknown, poss. form of *M. serriale* (below) ©JJ.

Pele's Murex *Homalocantha pele* WCP: Japan to New Caledonia, Marshalls and Hawaii, 60 mm. ©JJ.

Serriale Murex *Maculotriton serriale* IP: Red Sea, E. Africa to Marshalls and Marquesas, 24 mm. Elongate shell is white in color with dark brown markings. ©JJ.

Belligerent Rock Shell *Mancinella armigera* IP: E. Africa to Japan and Marshalls, 106 mm. Spire high, aperture white, with yellow or brown border. Photos (left, right): ©SJ, ©JJ.

Tuberose Rock Shell *Menathais tuberosa* IP: Red Sea, East Africa to Hawaii and Polynesia, 50 mm. Similar to M. armigera, but differs in having short spire and dark blotches on aperture margin. Photos (left, right): ©SJ, ©JJ.

Aglaos Rock Shell *Morula aglaos* WCP: Philippines, Marshalls, 14 mm. High spired shell, white with purple tinge, spiral rows of dark tipped tubercles and purplish aperture. Animal with black and yellow mottling. ©JJ.

Black&White Rock Shell *Morula albanigra* WCP: Japan to Guam and Marshalls, 7,5 mm. Small shell is white in color with dark brown bands. Animal white with black blotches. ©JJ.

Ambrosia Rock Shell *Morula ambrosia* WCP: Philippines to Marshalls, 21 mm. High spired shell, white in color, with mauve tinge, spiral rows of whitish tubercles and mauve aperture. Animal white with pinkish marbling. ©JJ.

Rough Rock Shell *Morula aspera* IP: Reunion to Japan and Marshall Islands, 17 mm. High spired shell, white in color, knobby with white nodules and dark interstices. Aperture purplish. Animal black with tan mottling. ©JJ.

Biconical Rock Shell *Morula biconica* IP: Reunion to Australia, Philippines and Marshalls, 30 mm. Elongate biconical shell with rows of white nodules and dark spiral bands between. Animal black with tan mottling. ©JJ.

Prickly Rock Shell *Morula echinata* IP: E. Africa to Hawaii, 10 mm. Creamy shell with orange nodes. ©JJ.

Striate Rock Shell *Morula striata* IP: Madagascar to Japan, Kiribati and Polynesia, 20 mm.

Grape Drupe *Morula uva* IP: Mauritius, Seychelles, to Tahiti, Marshalls, 30 mm. Shell spindle shaped with spinose tubercles. White in color, nodules dark brown or black, aperture light violet to purplish black. ©JJ.

Venus Comb Murex *Murex pecten* IP: E. Africa to Japan, Australia and Melanesia, 190 mm. Shell creamy white to tan in color. About 12 primary spines. Siphonal canal with 3 rows of spines, diminishing in size. ©JJ

Forskoehl' Murex *Murex forskoehlii* IO: Red Sea, also Mediterranean, Atlantic, 115 mm. © I. Sevriuk.

Thin Beak Murex *Murex tenuirostrum* IWP: Japan to Australia and New Caledonia, Solomons and Fiji, 140 mm. Brown with lighter colored spiral cords. Animal pale brownish with darker brown mottling.

Triple Spined Murex *Murex ternispina* IWP: India to Philippines, 150 mm. Creamy in color, aperture white.

Rare-Spined Murex *Murex trapa* IWP: E. Africa to Fiji, 124 mm. Light brown, aperture white . © Ria Tan.

Bramble Murex *Murex tribulus* IP: Red Sea, E. Africa, Indonesia to Philippines, 160 mm. White to light brown. *M. tribulus, M. ternispina* and *M. tenuirostrum* are hard to tell apart by photo. Photos (left, right): ©SJ, ©JJ.

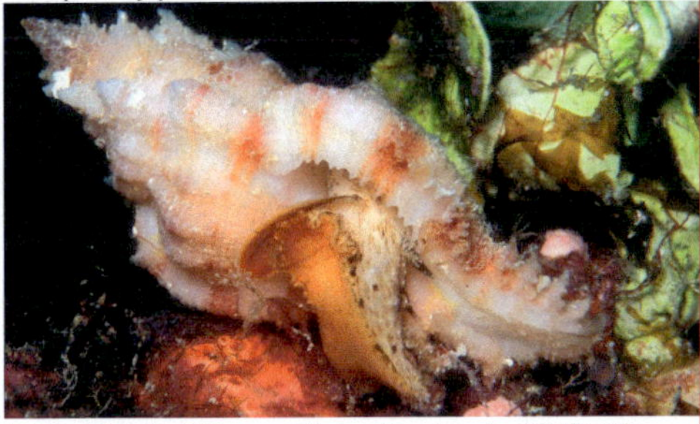

Woodcock Murex *Haustellum haustellum* IWP: Red Sea, East Africa to Australia and Fiji, 185 mm. ©SJ.

Three Angled Murex *Naquetia triqueter* IP: E. Africa to Marshalls, 75 mm. Creamy with brown bands. ©JJ.

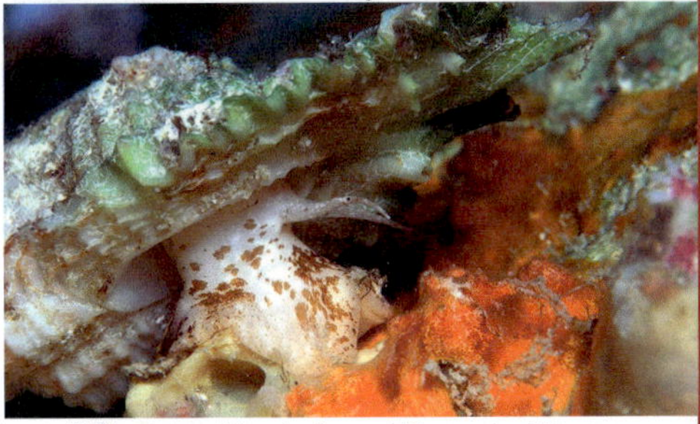

Cuming's Murex *Naquetia cumingii* IP: E. Africa to Maldives, Philippines, Marshalls and Tuamotu, 70 mm. Creamy shell w/brown bands, 2-3 on body. Close to *N. triqueter* in size and colors. Photos (left, right): ©SJ, ©JJ.

Wreath Jopas *Nassa serta* IP: Red Sea, E. Africa, to Maldives, Indonesia and Marshalls, 70 mm. Chestnut brown with white irregular blotches, aperture white. Animal black with white speckles. Photo on the right ©JJ.

Muricate Rock Shell *Pascula muricata* IP: Red Sea to Marshalls and Hawaii, 15 mm. Shell is white in color with high spire and strong ribs, upturned siphonal canal. Animal white with green network.

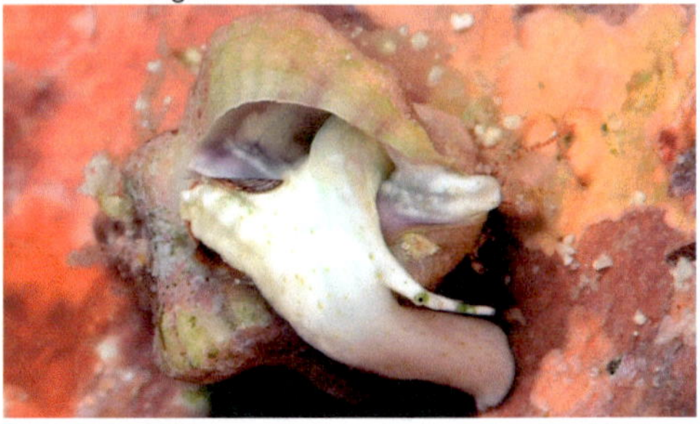

Purple-Mouth Rock Shell *Pascula ozenneana* IP: E Africa to Marshalls and Polynesia, 22 mm. White fusiform shell, aperture margins purplish. Animal white with green tinge, tubercles with green subapical band. ©JJ.

Sculptured Rock Shell *Phrygiomurex sculptilis* IP: E. Africa to Philippines and Marshall Islands, 22 mm. ©JJ.

Rudder Murex *Pterynotus elongatus* IP: Red Sea, E. Africa to Japan and Hawaii, 100 mm. © B. Abela.

Boutet's Murex *Pterynotus bouteti* WCP: N. Caledonia to Guam, Polynesia, 40 mm. White shell, pale-purple aperture. Animal white salmon-pink to whitish, covered with white circles. Photo (left, right): ©Bob Abela, ©JJ.

Fenestrate Murex *Pterynotus martinetanus* IP: Red Sea to Japan, Marshalls and Hawaii, 60 mm. White shell, surface with a reticulate appearance. The aperture is golden yellow, animal salmon-pink with white circles. ©JJ.

Three-Winged Murex *Pterynotus tripterus* IP: Red Sea, to Philippines and Marshalls, 60 mm. ©JJ.

Bubble Turnip *Rapa rapa* IP: E. Africa to Marshalls, 113 mm. Feeds on *Sarcophyton* leather corals. ©JJ.

Convoluted False Triton *Phyllocoma convoluta* IP: Red Sea, East Africa to Philippines and Marshalls, 30 mm. This white shell resembles a miniature version of tritons, was originally described under genus Triton. ©JJ.

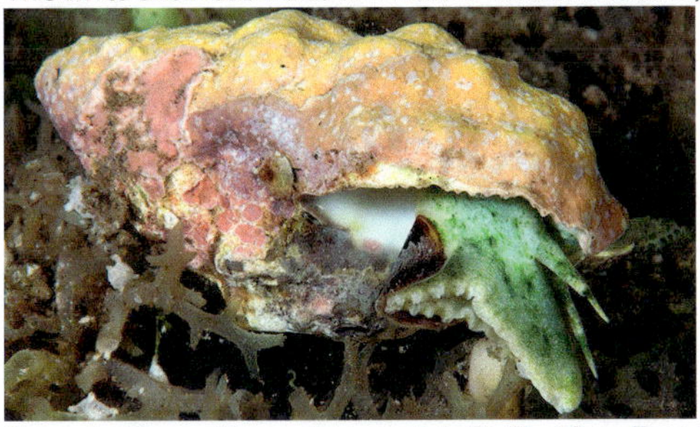

Granular Drupe *Tenguella granulata* IP: Red Sea, East Africa, Maldives to Marshalls, 30 mm. Shell with rectangular tubercles, dark grey to black, tubercles on white back ground, aperture violet. Animal greenish. Right photo: ©JJ.

Spiny Rock Shell *Spinidrupa euracantha* IP: Red Sea, E. Africa to Philippines and Marshalls, 25 mm. ©JJ.

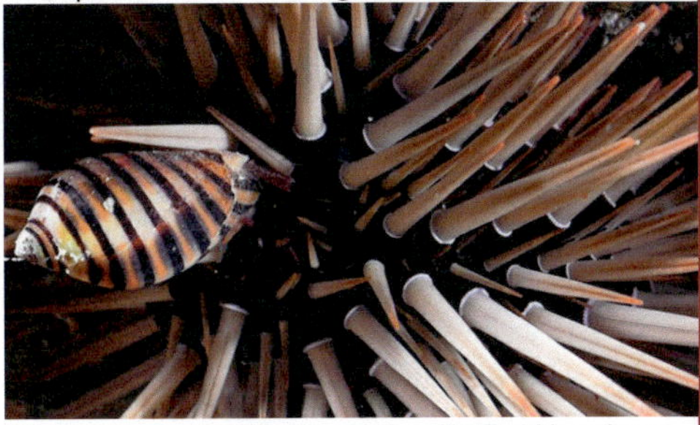

Ribboned Jopas *Vexilla vexillum* IP: E. Africa, Japan to Australia, Marshalls, 25 mm. © Shawn Miller.

Ribboned Jopas *Vexilla vexillum* (cont'd) Smooth shell, yellow-brown to cream in color, with brown spiral bands, 8 on body whorl, 2-3 on spiral whorls, aperture white, animal brown. Ectoparasite on sea urchins. ©JJ.

Minor Harp *Harpa amouretta* IP: Red Sea, E. Africa to Hawaii, 65 mm. Small, elongated shell, white in color with cream and brown banding. In danger, the animal can cast off the hind part of the foot. Photo (left, right): ©SJ, ©JJ.

Minor Harp *Harpa amouretta* (cont'd) unusual nearly white animal, Marshall Islands. ©JJ.

Articulate Harp *Harpa articularis* IWP: Burma, Thailand to Fiji, N. Caledonia and Japan, 110 mm.

Articulate Harp *Harpa articularis* (cont'd) Distinguished by narrow ribs and dark blotch, that nearly covers the ventral side of body whorl. Preys on crustaceans, trapping them by foot (left photo). Right photo: mating.

True Harp *Harpa harpa* IP: Red Sea, E. Africa to Hawaii and Polynesia, 110 mm. Distinguished by rectangular red or brown spots between ribs, the ventral side of body whorl with 3 separate brown blotches. ©JJ (left photo).

Kajiyam's Harp *Harpa kajiyamai* IWP: E. Africa to Japan, Philippines, 72 mm. © E.G de Suduiraut.

Imperial Harp *Harpa costata* IO: Mauritius; Reunion; Madagascar, 110 mm. © Charles E. Rawlings.

Major Harp *Harpa major* IP: Red Sea, E. Africa to Polynesia and Hawaii, 130 mm. Distinguished by the dark blotch, divided in the middle into 2 parts, on ventral side of body whorl. ©JJ.

MacDonald's Morum *Morum macdonaldi* CP: Marshall Islands endemic, 20 mm. ©SJ.

Ichiyama's Morum *Morum uchiyamai* WP, 65 mm. MNHN Kanacono 2016. © David Massemin.

Teramachi's Morum *Morum teramachii* Depth >90 m. WP: Japan to N.Caledonia, 66 mm. © E.G de Suduiraut.

Watanabe's Morum *Morum watanabei* WP: Japan to Philippines, 45 mm. Deep water. © E.G de Suduiraut.

Striped Olive *Miniaceoliva efasciata* WCP: N. Guinea to Polynesia, 75 mm. Cream-yellow shell with lilac flames and lines in irregular zigzag pattern, 2 interrupted transverse bands. Foot whitish with opaque white spots. ©JJ.

Head Concave Olive *Oliva concavospira* WP: Vietnam to Philippines and Japan, 44 m. Grey to brownish, with 198 pattern of longitudinal dark zigzag lines. © Philippe & Guido Poppe www.poppe-images.com.

Carnelian Olive *Oliva carneola* IP: Reunion to Micronesia and Polynesia, 28 mm. Small and broad cylindrical shell, spire covered by enamel, light orange in color with white and or brown bands. Right photo ©JJ.

Amethyst Olive *Oliva amethystina* IP: E. Africa to Philippines, Indonesia, Marshalls, 52 mm. *O. amethystina* and *O. mantichora* were for centuries confused under the name *O. annulata,* now a *nomen dubium*. ©JJ.

Amethyst Olive *Oliva amethystina* (cont'd) Shell with a rather well elevated spire and dark brown spots below.

Irisans Olive *Oliva irisans* IP: Mauritius to Philippines, Australia, New Caledonia and Fiji, 50 mm.

Irisans Olive *Oliva irisans* (cont'd) The shell is white or cream in color, with a pattern of brown tents and longitudinal zigzag lines. The spire is low. Photo on the right - *Oliva irisans* f. *concinna*

Keen's Oliva *Oliva keenii* WP: Philippines and Indonesia to New Guinea, 55 mm. The shell is greenish or creambrown in color, with a pattern of brown longitudinal zigzag lines. The spire is low. Right photo ©JJ.

Lambert's Oliva *Oliva lamberti* WP: Australia to New Caledonia, Fiji, Vanuatu, 75 mm. Interior pale orange.

False Amethyst Olive *Oliva mantichora* IWP, 53 mm. © Philippe & Guido Poppe www.poppe-images.com.

Red-Mouth Olive *Oliva miniacea* IWP, 100 mm. Large shell, cylindrical with a relatively low, pointed spire, interior orange in color. Outside creamy white, with wavy axial lines and 2-3 spiral bands of brown blotches.

Weasel Olive *Oliva mustelina* IWP: Madagascar to Japan, Singapore and New Guinea, 49 mm. © Ria Tan.

Peg Olive *Oliva ozodona nitidula* IP: East Africa to the Philippines and Micronesia, 27 mm. ©JJ.

Silk-Clad Olive *Oliva panniculata* IP: E. Africa, Madagascar to Marshalls and Polynesia, 25 mm. Small shell, white in color, with spiral bands of brown spots or wavy axial red-brown lines. Animal white. ©JJ.

Ponderous Olive *Oliva ponderosa* IWP, 73 mm. © Philippe & Guido Poppe www.poppe-images.com.

Chevron Olive *Oliva rufula* WP: Indonesia, Philippines to Solomons, 40 mm. Grey or brown with tent-marks.

Blood Olive *Oliva reticulata* IP: E. Africa to Japan, New Caledonia and Polynesia, 50 mm. Low spire orange columellar callus. Usually a longitudinal pattern of small tent-marks or zigzag lines and 2 broad dark bands.

Tiger Olive *Oliva tigridella* WCP, 42 mm. Cylindrical shell with high spire, white to creamy in color, some with a pattern of large dark brown or black blotchy bars. © Ria Tan.

Tricolor Olive *Oliva tricolor* IWP, 63 mm. Low spire with oblique, orange-tinged dark lines, outer color a mix of green, blue, orange spots. Photo on the right: preying on the cardinal fish.

Black Olive *Oliva vidua* IP, 70 mm. Shell swollen posteriorly, spire very low or depressed. ©JJ.

Richert's Oliva *Omogymna richerti* CP: Hawaii, 35 mm. Shell swollen posteriorly, high spired. ©SJ.

Montrouzier's Ancilla *Amalda montrouzieri* WP: New Caledonia, 42 mm. © David Massemin.

Amon's Olivellopsis *Olivellopsis amoni* WCP: 7 mm. N. Guinea, Indonesia, Marshalls. © C.Sylvester.

Red Mitre *Atlantilux rubra* IP: Reunion to Marshalls, 10 mm. Small shell, pink to cherry red in color. ©JJ.

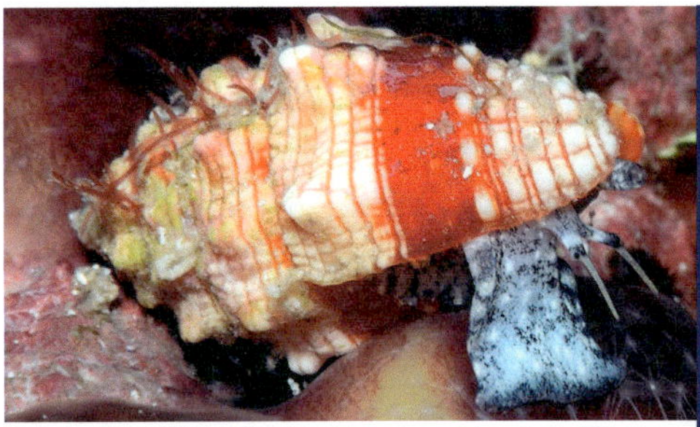

Patriarchal Mitre *Orphanopusia patriarchalis* IP: Seychelles to Japan, Australia and Hawaii, 30 mm.

Two-Lined Mitre *Protoelongata bilineata* WP: the Philippines to Guam, 16 mm. © Bob Abela.

Coral Mitre *Protoelongata corallina* WP: Philippines to Fiji, 16 mm. Small shell, orange or red in color. ©SJ.

Helene's Mitre *Protoelongata heleneae* CP: French Polynesia, Marshall Islands, 21 mm. ©JJ.

Loyalty Islands Mitre *Protoelongata loyaltyensis* WCP: Japan to New Caledonia and Marshalls, 15 mm. ©SJ.

Red-Ribbon Mitre *Protoelongata rubrotaeniata* CP: Solomons to Marshalls, 17 mm. ©JJ.

Lauta Mitre *Pusia lauta* WCP: Japan to Philippines, N. Guinea, Guam and Hawaii, 15 mm. © D. Burdick.

Abela's Mitre *Thala abelai* WP: Guam, Marshalls, Vanuatu, N, Caledonia, 11 mm. © Bob Abela.

Cernica Mitre *Thala cernica* IWP: Madagascar and Mauritius to Guam, 12 mm. © Bob Abela.

Exquisite Mitre *Thala exquisita* CP: Marshall Islands, Tuamotu, 14 mm. Orange to pink with 8-9 whorls. ©JJ.

Lilico's Mitre *Thala lillicoi* CP: Marshalls, Hawaii, 6 mm. Purple shell, Animal with yellow spots. ©JJ.

Gori's Mitre *Thala gorii* IP, 8 mm. Differs from *T.jaculanda* by lower spire, fewer spiral cords. © Bob Abela.

Jaculanda Mitre *Thala jaculanda* IWP: S. Africa and Oman to Guam, 9 mm. © Bob Abela.

Pinpricked Mitre *Vexillum acupictum* IP: Red Sea, E. Africa to Japan and Marshalls, 46 mm. ©JJ.

Slender Mitre *Vexillum angustissimum* IP: Red Sea to Philippines, Fiji, N. Zealand, 20 mm. ©SJ.

Lovely Mitre *Vexillum amabile* IP: Red Sea, E. Africa to Japan, N. Caledonia and Marshalls, 20 mm. ©JJ.

Small Girdle Mitre *Vexillum balteolatum* WP, 65 mm. © Philippe & Guido Poppe www.poppe-images.com.

Golden Mitre *Vexillum aureolatum* IP: Red Sea, E. Africa to Hawaii and Marshalls, 25 mm. Shell relatively broad, axial ribs strong. Red-orange in color with white transverse bands. Animal grey with white dots. ©JJ.

Golden-Lined Mitre *Vexillum aureolineatum* WCP: Australia to Philippines, and Hawaii, 20 mm. Inflated shell with recurved siphon canal, orange-yellow in color with red transverse lines. ©JJ.

Two-Part Mitre *Vexillum bipartitum* IP: Red Sea, East Africa to Philippines and Marshalls, 7 mm. ©JJ.

Two-Zone Mitre *Vexillum bizonale* IP: Red Sea, East Africa to Philippines, Marshalls, 20 mm. ©JJ.

Colorless Mitre *Vexillum discolorium* IWP: Japan to Indonesia, Solomons, Guam, 15 mm. ©SJ.

Cadaverous Mitre *Vexillum cadaverosum* IP: E. Africa to Philippines, Hawaii and Marshalls, 25 mm. ©JJ.

Kaffraria Mitre *Vexillum caffrum* IP: E. Africa to Philippines, Australia and Solomons, 60 mm. ©SJ.

Regal Mitre *Vexillum citrinum* IP: East Africa to Philippines, Indonesia and Solomons, 90 mm. ©SJ.

Bloody Mitre *Vexillum consanguineum* IP: Seychelles to Philippines, Marshall Islands, 15 mm. ©JJ.

Little Basket Mitre *Vexillum corbicula* IP: Mauritius, Philippines to Hawaii, 20 mm. © Bob Abela.

Cancellaria Mitre *Vexillum cancellarioides* IP: Red Sea, E. Africa to Japan, Marshalls and Hawaii, 25 mm. White, broad in shape shell with brown transverse band on the body whorl. Animal white. ©JJ.

Garland Mitre *Vexillum coronatum* IP: Red Sea, Maldives to Philippines and Marshalls, 25 mm. ©JJ.

Taki's Mitre *Vexillum costatum* IP: E. Africa to Japan, Australia, Tuamotu and Marshalls, 63 mm. ©SJ.

Saffron Mitre *Vexillum crocatum* IP: Red Sea to Philippines, Marshalls, Hawaii, 36 mm. Shell relatively broad in shape, orange to brown, with white band on the body whorl. Animal with white and brownish speckles. ©JJ.

Saffron Mitre *Vexillum crocatum* f. *cumingi*. Former *Vexillum cumingi* is now treated as *Vexillum crocatum*. Differs in having dark interstices between spiral cords. ©JJ.

Red-Front Mitre *Vexillum diutenerum* IP: E. Africa to Philippines, Marshalls and Hawaii, 10 mm. ©JJ.

Roughened Mitre *Vexillum exasperatum* IP: E. Africa to Maldives, Japan and Marshalls, 29 mm. ©JJ.

Spiny Mitre *Vexillum echinatum* IP: Red Sea to Philippines, Tahiti and Marshalls, 30 mm. Elongate, turreted white shell with a red-brown transverse band, axial ribs with spinose nodules. ©JJ.

Scarlet Mitre *Vexillum coccineum* WP: Thailand to Japan, Philippines and Indonesia. 87 mm. Elongate fusiform shell, red-orange in color, decorated with yellow transverse bands. Animal white with grey speckling. ©JJ.

Excellent Mitre *Vexillum exquisitum* IP: E. Africa, Japan to Philippines, Marshalls and Polynesia, 10 mm. Small biconic shell with strong axial ribs. Red with a pattern of transverse wide bands and dark-brown lines. ©JJ.

Fidicula Mitre *Vexillum fidicula* WP, 20 mm.
MNHN Koumac 2018. © David Massemin.

Firmly Folded Mitre *Vexillum fortiplicatum*
ZIP: East Africa to Philippines and Hawaii, 12 mm.

Gage's Mitre *Vexillum gagei* CP: Marshall Islands, Hawaii, 15 mm. ©SJ.

Goubini Mitre *Vexillum goubini* IP: E. Africa to Japan, Guam and Tuamotu, 8 mm. © Bob Abela.

Granular Mitre *Vexillum granosum* WCP: Philippines to N. Caledonia, Palau and Marshalls, 52 mm. ©JJ.

Low Mitre *Vexillum humile* IP: E. Africa to Maldives, Philippines and Marshalls, 15 mm. ©JJ.

Gruner's Mitre *Vexillum gruneri* IP: E. Africa, Philippines to N. Guinea, N. Caledonia and Marshalls, 60 mm. Shell pinkish-grey to creamy with two to three reddish-brown interrupted spiral threads on the body whorl. ©JJ.

Unhappy Mitre *Vexillum infaustum* IP: E. Africa to Philippines, N. Guinea, Hawaii, 14 mm. ©JJ.

Interrupted Mitre *Vexillum interruptum* IP: Red Sea, Philippines to New Caledonia and Hawaii, 8 mm. ©JJ.

Interstriated Mitre *Vexillum interstriatum* IP: Reunion, Philippines to Australia, Hawaii, 47 mm. ©SJ.

Ribboned Mitre *Vexillum intertaeniatum* WP: Philippines to Guam, 26 mm. © Bob Abela.

Ismene Mitre *Vexillum ismene* WP, 17 mm. MNHN Koumac 2018. © David Massemin.

Johnson's Mitre *Vexillum johnsoni* WCP: Philippines to Marshalls, 12 mm. ©SJ.

Jeanetteae's Mitre *Vexillum jeanetteae* WCP: Philippines, Samoa, Marshalls, 15 mm. Shell inflated, biconic in shape. Yellow with transverse brown bands. Black animal with white dots. Photo: ©JJ.

Kathie Way's Mitre *Vexillum kathiewayae* WCP: New Guinea to Japan, Guam and Marshalls, 27 mm. ©JJ.

Kuiper's Mitre *Vexillum kuiperi* WCP: Japan to Guam, Marshalls, 11 mm. Inflated shell with red bands. ©JJ.

Lenhill's Mitre *Vexillum lenhilli* WCP: Philippines to Marshalls and Hawaii, 7 mm. Shell relatively broad in shape. Fawn to brown or orange, with transverse white and narrow black bands. Animal tan with white speckles. ©JJ.

White-Belt Mitre *Vexillum leucozonias* IP, 25 mm. Shell from fawn to dark-brown with white bands. ©JJ.

McCausland's Mitre *Vexillum mccauslandi* WP: Indonesia, 24 mm. Slender shell with brown stripes.

Bright Mitre *Vexillum lucidum* IP: Red Sea, E. Africa to Philippines, Australia and Marshalls, 20 mm. White to creamy in color, elongated shell. Pattern of spiral cords, not interrupted on axial ribs. ©JJ.

Michau Mitre *Vexillum michaui* IP: Red Sea, E. Africa to Philippines, Australia and Marshalls, 27 mm. Shell small, white in color, body whorl with 3-5 rows of orange spots on the axial ribs. ©JJ.

Tiny Mitre *Vexillum micra* IP: Reunion, Oman, Philippines to Australia, Guam and Hawaii, 10 mm. Small conical shell, yellow to dark brown in color, with transverse white interrupted bands. Photo (left, right): © B. Abela, ©JJ.

Thousand-Rib Mitre *Vexillum millecostatum* IP: E. Africa, Japan to Indonesia, Marshalls and Hawaii, 35 mm. Biconical shell, fawn to dark brown in color, with transverse yellow bands. Animal black with white dots. ©JJ.

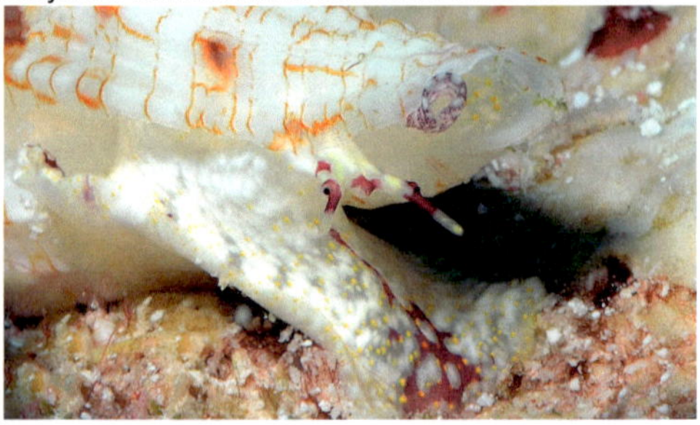

Miraculous Mitre *Vexillum mirabile* IP: Red Sea, Philippines to Australia and Marshalls, 66 mm. Shell slender in shape, white with irregular red-brown spots. Animal white with mauve reticulation and yellow dots. ©JJ.

Basket Mitre *Vexillum modestum* IP: East Africa, Philippines to Marshalls and Hawaii, 30 mm. ©JJ.

Multi-Marked Mitre *Vexillum plurinotatum* IP: E. Africa, Reunion, Japan and Marshall Islands, 7 mm. ©JJ.

Pagoda Mitre *Vexillum pagodula* WCP: Philippines to New Guinea, Fiji and Marshalls, 12 mm. Small shell, fawn to dark brown in color, with transverse white lines. Animal white with brownish blotches. ©JJ.

Paligerum Mitre *Vexillum paligerum* IP: Red Sea, Philippines to Marshalls, 14 mm. Shell creamy to red in color, with interrupted yellow transverse bands on the axial ribs. Photo on the left - juvenile, 7 mm. ©JJ.

Plicate Mitre *Vexillum plicarium* IP: Andaman Sea, Japan to Australia and Samoa, 66 mm. Biconical, white in color shell with black transverse bands. Animal grey with dark speckles. Photo (left, right): © R. Groeneveld, © U. Weinreich.

Polygonal Mitre *Vexillum polygonum* WP, 40 mm. © Philippe & Guido Poppe www.poppe-images.com.

Potier's Mitre *Vexillum potieri* IP: Reunion to Philippines, Solomons and Marshalls, 45 mm. ©SJ.

Root Mitre *Vexillum radix* WCP, 25 mm. Slender fawn to dark brown shell with white transverse bands.

Red-Ribbed Mitre *Vexillum rubrocostatum* WP: Philippines to Solomons and Fiji, 44 mm. ©SJ.

Recurved-Beak Mitre *Vexillum recurvirostris* WCP: Philippines to Guam, Marshalls, 15 mm. Shell small, variable in color: grey to yellow ocher and brown, with yellow ribs and diffuse white bands. ©JJ.

Bent-Beaked Mitre *Vexillum* sp. Marshalls, 14 mm. Undescribed species, close to *C. recurvirostris*. ©JJ.

Rose-Colored Mitre *Vexillum roseum* IP: Red Sea, Reunion, Indonesia to Guam, Marshalls, 13 mm.

High-Spired Mitre *Vexillum longispira* WCP, 22 mm. Rare species, close to *V. radius*. ©JJ.

Rugose Mitre *Vexillum rugosum* IP: 64 mm. © Philippe & Guido Poppe www.poppe-images.com.

Bloodsucker Mitre *Vexillum sanguisuga* IP: E. Africa and Thailand to Philippines, Australia, Fiji and Polynesia, 50 mm. Elongate shell, white to grey in color. Interior violet, edge of aperture with a broad purple zone. ©JJ.

Bloodsucker Mitre *Vexillum sanguisuga* (cont'd) Red dots on the summits of the axial ribs. Animal light grey, densely variegated with darker grey.

Elegant Mitre *Vexillum scitulum* IWP: East Africa, Madagascar, Philippines, 36 mm.

Engraved Mitre *Vexillum sculptile* IWP: Red Sea to Japan, Philippines and Fiji, 46 mm.

Half-Banded Mitre *Vexillum semifasciatum* WP: Japan to Truk, Guam, Fiji, 40 mm. © Rokus Groeneveld.

Half-Sculpted Mitre *Vexillum semisculptum* IWP: Andamans to Japan, Samoa and Fiji, 31 mm.

Siciliai Mitre *Vexillum siciliai* WCP: New Caledonia to Palau Islands and Micronesia, 26 mm. Elongate shell is brown in color, with light brown axial ribs and narrow white transverse bands. NHMN Koumac-2018 expedition.

Tapering Mitre *Vexillum* sp. Marshalls, 32 mm. White shell with faint yellow band on the body whorl. ©JJ.

Specious Mitre *Vexillum speciosum* IWP: Reunion to Chagos and Marshall Islands, 22 mm. ©JJ.

Herringbone Mitre *Vexillum spicatum* WP: Japan to Philippines and New Guinea, 31 mm. Elongate fusiform shell, beige to pale brown. Animal grey with yellow-black mottling.

Barb Mitre *Vexillum spiculum* IP: Madagascar, Marshall Islands, 15 mm. Rather broad shell, white in color, with faint red transverse reddish lines, often interrupted. Photo (left, right): ©SJ, ©JJ.

Stainforth's Mitre *Vexillum stainforthii* IP: Japan to Philippines, Guam, Australia. © Sven Kahlbrock.

Thorsson's Mitre *Vexillum thorssoni* WP: Philippines, 17 mm. Broad white shell with red or brown bands.

Turban Mitre *Vexillum turben* IP: E. Africa to Japan, Marshalls and Hawaii, 25 mm. Pyriform inflated shell, orange in color. Dark grey animal with white and yellow bands. ©JJ.

Turreted Mitre *Vexillum turriger* IP: Madagascar to N. Caledonia and Marshalls, 18 mm. Shell is white in color, with faint interrupted red band. Animal white with mauve reticulation anterior. ©JJ.

Golden Mitre *Vexillum unifasciale* IP: E. Africa to Japan, Philippines and Hawaii, 29 mm. Inflated rather broad shell, yellow to white in color, some with brown or red-orange transverse bands. ©JJ.

Turnip Mitre *Vexillum tusum* IP: Red Sea, Reunion to Philippines, Australia and Marshalls, 12 mm. Small biconic shell is brown to red in color, with white or alternative red-white transverse bands and blotches. ©JJ.

Obelisk Mitre *Vexillum obeliscus* IWP: Red Sea, E. Africa to Philippines and Australia, 45 mm. ©SJ.

Blunt-Spines Mitre *Vexillum* cf. *obtusispinosum* CP: Marshalls, 13 mm. Undescribed species. ©SJ.

Little Fox Mitre *Vexillum vulpecula* IWP: Japan to Philippines and Australia, 72 mm. Turreted shell with pointed spire. Variable in color, but generally orange in color, banded with white and black. NHMN Koumac-2018 expedition.

Decorated Mitre *Vexillum unifasciatum* IP: E. Africa to Japan, Australia, Marshalls and Hawaii, 72 mm. ©JJ.

Zebuense Mitre *Vexillum zebuense* IWP: Thailand to Japan, Indonesia, Australia, N. Caledonia, 42 mm.

Jealous Mitre *Vexillum zelotypum* WCP: Japan and Korea to Fiji, Marshall Islands, 21 mm. Fusiform shell, strong axial ribs over-ridden by spiral riblets, creating a checkered pattern. ©JJ.

TURBINELLIDAE

Ceram Vase *Vasum ceramicum* IP: E. Africa to Japan, Australia, Marshalls and Polynesia, 16 cm. Spire relatively high, 3 equal-sized columellar folds. Feeds on eunicid polychaetes. Photo (left, right): ©SJ, ©JJ.

Top Vase *Vasum turbinellus* IP: Red Sea, E. Africa to Japan, Australia and Polynesia, 16 cm. Spire relatively low; 4 or 5 unequal columellar folds. Feeds on polychaete annelids and sipunculans. Photo on the right ©JJ.

CYSTISCIDAE

Ovate Gibberula *Gibberula ovata* WCP: Japan to Marshalls, 6 mm. Nocturnal, hides during the day, parasitizing sleeping fish at night. Left: ©JJ. Right: feeding on *Scarus frenatus* parrotfish - ©SJ. ID tentative.

Elongated Gibberula *Gibberula* sp. Marshall Islands, 3.2 mm. ID tentative. ©JJ.

White Gibberula *Gibberula* sp. Marshall Islands, 3 mm. © Christina Sylvester. ID tentative.

Tiger Cystiscus *Cystiscus* sp. WCP: Indonesia, Philippines to Marshalls, 4 mm. ©JJ.

Mariano's Crithe *Crithe* aff. *marianoi* WP: Indonesia, Philippines to Fiji, 4 mm. ©JJ.

Oral Marginella *Dentimargo ringicula* IP: Madagascar, Marshalls, 2.3 mm. ©JJ. ID tentative.

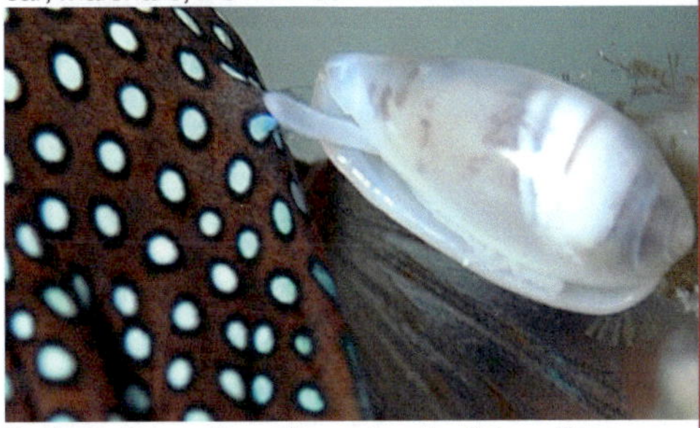

Sagami Marginella *Hyalina sagamiensis* WP: Japan, Philippines, 9 mm. Parasitizes sleeping fishes, like some other family members. © Philippe & Guido Poppe www.poppe-images.com.

Philippine Marginella *Volvarina philippinarum* WCP: Philippines to Australia and Marshalls, 16 mm. Smooth and polished elongate shell, white in color. Animal white with orange spots. ©JJ. ID tentative.

Mottley Marginella *Volvarina* sp. CP: Marshall Islands, 14 mm. Fusiform elongate shell. Animal translucent, with opaque white spots and red dots. ©JJ.

Banded Marginella *Volvarina* sp. CP: Marshall Islands, 3 mm. © Christina Sylvester. ID tentative.

Beige Marginella *Volvarina* sp. WP: Guam, 5 mm. © Bob Abela. ID tentative.

Amazing Nutmeg Shell *Plesiotriton mirabilis* 24 mm. WCP: Guam, Marshalls, 24 mm. © Bob Abela.

Ladder Nutmeh Shell *Scalptia scalarina* IWP: East Africa, India to Thailand, Japan, Australia, 27 mm.

Tabulate Nutmeg Shell *Scalptia contabulata* IWP: E. Africa to Philippines and Fiji, 30 mm. Shell is cream to brownish, with prominent axial ribs and brown transverse bands. Left: © D. Massemin, MNHN Koumac 2018.

Twisted Nutmeg Shell *Scalptia crenifera* IWP: India to Japan, Philippines and Australia, 30 mm.

Oblique Nutmeg Shell *Scalptia obliquata* IWP: E. Africa to Japan and New Caledonia, 30 mm. ©JJ.

Orange Nutmeg Shell *Scalptia* sp. WP: Indonesia, 16 mm. Orange shell with widely spaced axial ribs. ©JJ.

Mercado's Nutmeg *Scalptia mercadoi* WP: Philippines, 36 mm, depth 80-120 m. © E.G de Suduiraut.

Thysthlon's Nutmeg *Trigonostoma thysthlon* WP: Japan- Australia. Depth 150 m. © E.G de Suduiraut.

Antiquated Nutmeg Shell *Tritonoharpa antiquata* IP: Red Sea, Japan, Australia to Marshalls, 65 mm. ©JJ.

Deshayes's Volute *Cymbiola deshayesi* SP: New Caledonia, 105 mm. The shell is fusiform, solid, with conic spire. White to cream with 2 broad interrupted red bands. Animal red with yellow dots, foot with yellow margin.

Noble Volute *Cymbiola nobilis* WP: Thailand to Singapore and Philippines, 222 mm. Fusiform shell, beige with a zig-zag lines. Animal black with yellow spots. Right: © Arthur Anker, juvenile. Left: © Tsu Soo Tan.

Palawan Volute *Cymbiola palawanica* WP: Philippines, Palawan Island, 110 mm. © B. Jean-Pierre.

Rossini's Volute *Cymbiola rossiniana* SP: New Caledonia (endemic), 185 mm. © Luc Faucompre.

Bat Volute *Cymbiola vespertilio* WP: Thailand to Vietnam, Philippines, Indonesia and Australia, 160 mm. Outer coloration variable, generally cream to olive brown, with darker overlays of zigzag lines, blotches or streaks.

Vexillate Volute *Harpulina arausiaca* IO: Sri Lanka, India, 105 mm. © Charles E. Rawlings.

Brown-Lined Volute *Harpulina lapponica* IO: Sri Lanka, India, 103 mm. ©: Charles E. Rawlings.

Clover's Lyria *Lyria cloveriana* IO: Sri Lanka, Thailand, 98 mm. © Charles E. Rawlings.

Delightful Lyria *Lyria deliciosa* SP: Australia, New Caledonia, 42 mm. © Uwe Weinreich.

Bohol's Lyria *Lyria boholensis* WP: Philippines, New Caledonia, Borneo, 86 mm. © E.G de Suduiraut.

Diadem Volute *Melo amphora* WP, 550 mm. Shoulder with elevated spines. © Uwe Weinreich.

Crowned Baler *Melo aethiopicus* WP: Indonesia to Vietnam and Philippines, Malaysia, N. Guinea, 250 mm. Globose shell, red-orange in color, with low spire. Shoulder with short elevated spines. © Ludovic David.

Crowned Baler *Melo broderipii* WP: Indonesia to Philippines and N. Guinea, 360 mm. Globose shell, cream to pale brown in color. Shoulder with short elevated spines. © Philippe & Guido Poppe www.poppe-images.com.

Cymbium Melo *Melo melo* WP: India to Philippines and Indonesia, 365 mm. Large globose shell, spire covered by body whorl, shoulder devoid of spines. Outside orange, often with banding of brown spots. © Ria Tan.

Nigrita Nerite *Nerita nigrita* WP: Malaysia, Indonesia, 30 mm. Shell with spire depressed, with fine spiral ribbing. Exterior coloration variable, generally olive-grey streaked or banded with black and white.

Polished Nerite *Nerita polita* IWP: Red Sea, East Africa to Philippines, Australia, Hawaii, 40 mm.

Emerald Nerite *Smaragdia rangiana* IWP: N. Guinea, Madagascar, Mauritius to Philippines, 11 mm.

Emerald Nerite *Smaragdia viridis* IP: Red Sea, Philippines, also Mediterranean, Atlantic, 10 mm. The shell is oval, depressed, smooth and glossy. Lives on seagrass intertidally and in the shallow subtidal.

Pinkish Cap Shell *Plesiothyreus* sp. CP: Marshalls to French Polynesia, 10 mm. Pink or reddish, with granulose ribbing. Photo: (left, right): © C. Sylvester, © Arthur Anker.

Trap-Door Nerite *Neritopsis radula* IP: East Africa to Japan, Australia, Marshalls, 35 mm. ©SJ.

Shell-Less Nerite *Titiscania* sp. Indonesia, 14 mm. © Arthur Anker. ID tentative, close to *T. limacina*.

NERITOPSIDAE

Slug-Like Titiscania *Titiscania limacina* IP: Red Sea, Mauritius to Philippines, Australia, Marshall Islands, 18 mm. Shell-less neritoidean snail with slug-like appearance. Found under intertidal rocks. ©JJ.

CALLIOSTOMATIDAE

Ticao Top Shell *Calliostoma ticaonicum* WCP: Japan to Philippines and Marshalls, 20 mm. Perforate conical shell, yellow-brown to pale pink in color, aperture is white inside. ©JJ.

Caledonian Top Shell *Calliostoma* sp. SP: New Caledonia, 10 mm. MNHN Koumac 2019. © D. Massemin.

Beautiful Top Shell *Coralastele pulcherrima* WP: Philippines, Indonesia, 10 mm.

Strawberry Top Shell *Calthalotia fragum* WCP: Philippines to Australia and Marshall Islands, 24 mm. Glossy conical shell with pattern of orange strokes on spiral cords. Animal white with fine red lines. ©JJ.

PHASIANELLIDAE

Solid Pheasant *Phasianella solida* IP: Red Sea, E. Africa to Japan, Australia and Marshalls, 35 mm. Long, thin and turban-shaped shell with 4-5 rounded whorls. Coloration variable: red to olive. Photo (left, right): ©SJ, ©JJ.

Cone-Shaped Top Shell *Rochia conus* IP: Red Sea to Vietnam, Philippines and Marshalls, 80 mm. Conical shell, longer than wide. Variable in color, surface of body whorl with beaded spiral threads. Photo on the right ©JJ.

Commercial Top Shell *Rochia nilotica* IP: East Africa to Australia, N. Caledonia, Marshalls and Polynesia, 165 mm. Large conical shell, white, with large, irregularly axial reddish stripes. Economically important. ©JJ.

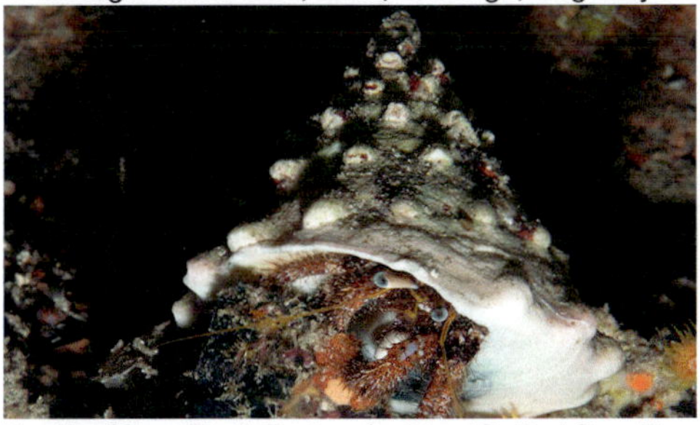

Toothed Top Shell *Tectus dentatus* IO: Red Sea, Persian Gulf, 151 mm. Shell with rounded protuberances.

Striped Top Shell *Tectus virgatus* IO: Red Sea to Madagascar, Mauritius, 45 mm.

Pyramid Top Shell *Tectus pyramis* IP: Sri Lanka to Japan, New Caledonia and Polynesia, 15 cm.

Fenestrate Top Shell *Tectus fenestratus* IP: Malaysia to Philippines, Australia, Marshalls, 70 mm. ©JJ.

Tall Top Shell *Tectus triserialis* WCP: Malaysia to Japan, Indonesia, Australia, Marshalls, 60 mm. Conical shell, longer than wide, flesh-colored. The base of the shell is nearly flat. Photo (left, right): ©SJ,©JJ.

Rosemouth Star Shell *Astralium rhodostomum* IP: Mauritius to Maldives, Japan, Marshalls and Tahiti, 55 mm. Conical shell, reddish in color, solid superior nodules near to the suture. Photo: (left, right) ©SJ, ©JJ.

Silvermouth Turban *Turbo argyrostomus* IP: E. Africa to Japan, Australia, Polynesia and Marshalls, 100 mm. Large turbinate shell, color of the aperture mostly silvery white. Feeds on algae. Used as food.

Goldmouth Turban *Turbo chrysostomus* IP: E. Africa to Japan, Australia, N. Caledonia and Melanesia, 80 mm. Large turbinate shell, color of the aperture mostly silvery white. Aperture orange to yellow, with white margins.

Tapestry Turban *Turbo petholatus* IP: Red Sea, E. Africa to Japan, Australia, 85 mm. Smooth polished shell, variable in color and pattern, generally red-brown to greenish with dark spiral bands. Right photo ©JJ.

Rough Turban *Turbo setosus* IP: Madagascar to Japan, N. Caledonia, Marshalls and Polynesia, 80 mm. Shell outside with strong spiral cords, irregularly maculated with dark brown. Outer lip crenulated, aperture white. ©JJ.

Orange Mouthed Top Shell *Chrysostoma paradoxum* WP: Australia to Philippines, 25 mm. © Ria Tan.

Secret Clanculus *Clanculus clanguloides* IP: East Africa to Australia, Japan and Marshalls, 12 mm. ©SJ.

Beautiful Clanculus *Clanculus margaritarius* IP: Reunion to Japan, Australia, Samoa and Marshalls, 20 mm. Shell outside Reddish-brown in color, with strong spiral cords, irregularly maculated with black. ©JJ.

Strawberry Top Shell *Clanculus pharaonius* IO: Red Sea, Gulf of Oman, 25 mm.

Purplish Clanculus *Clanculus puniceus* IO: Red Sea, East and South Africa, 22 mm. © Rafi Amar.

Purple Clanculus *Clanculus samoensis* WCP: Japan to Australia and Marshalls, 13 mm. Small shell with purple protoconch and strong beaded spiral cords Animal white with purplish mottling. ©JJ.

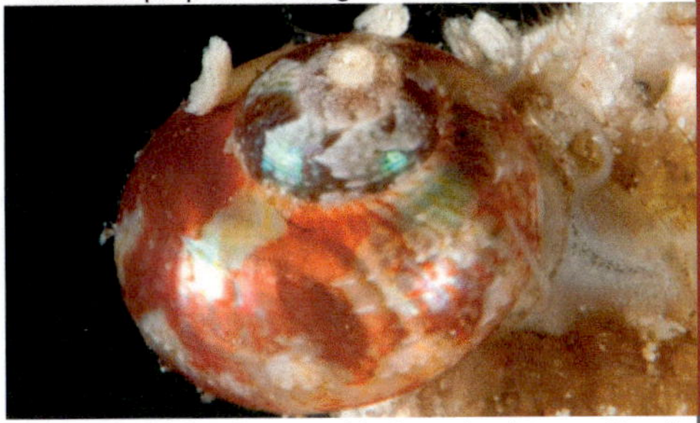

Stearns' Top Shell *Ethminolia stearnsii* IWP: E. Africa to Japan, Philippines and Indonesia, 10 mm. Small, depressed, shining shell, vermiculate pattern with darker clouds and spaced spiral lines.

Guam Button Top Shell *Ethalia guamensis* WP: Japan, Philippines, Indonesia, Australia, 25 mm.

Gilbert's Top Shell *Jujubinus polychroma* WCP: Philippines to Australia and Marshalls, 20 mm.

Gilbert's Top Shell *Jujubinus polychroma* (cont'd) Conical shell, longer than wide. Variable in color, red to green, spiral cords with white maculations and oblique axial lines. ©JJ.

Mottled Top Shell *Jujubinus* sp. CP: Marshalls, 15 mm. Close to *J. polychroma* but shell sculpture differs. ©JJ.

Labio Top Shell *Monodonta labio* IWP: Reunion to Philippines and Australia, 45 mm. © Ria Tan.

Belcher's Top Shell *Monilea belcheri* IWP: Andaman Sea to Japan, Philippines and Australia, 17 mm. Depressed conoidal shell, variable in color: grey-brown to reddish with black maculations on spiral cords.

Polished Top Shell *Sericominolia vernicosa* WP: Japan to Philippines, Fiji, Samoa and Australia, 12 mm. Depressed globose shell, variable in color from white to fawn and red. Left: © Bob Abela.

Papery Stomatella *Pseudostomatella papyracea* IP: E. Africa to Philippines, Indonesia, New Guinea and Hawaii, 32 mm. Globose shell with shining, polished surface, variable in color. Nocturnal, fast moving animal.

Papery Stomatella *Pseudostomatella papyracea* (cont'd) Shell is fawn colored, with a series of short markings at the periphery alternately reddish and white. Two white-edged rounded flaps between the tentacles.

Nucleus Top Shell *Rossiteria nucleus* IWP: Madagascar to Japan, Philippines and New Caledonia, 11 mm.

Maculated Top Shell *Trochus maculatus* IP, 70 mm. Whitish with axial stripes of red or brown. ©JJ.

False Ear Shell *Stomatella auricula* IP: Red Sea to Japan, Indonesia, Marshalls, 20 mm. Widespread, but rarely noticed by divers, secretive, highly variable species. Nocturnal, hides under shallow rocks during the day.

False Ear Shell *Stomatella auricula* (cont'd) The color varies from red to white and even blackish, with more or less numerous white spots and often darker spiral bands.

Greenish False Ear Shell *Stomatella* cf. *auricula* WP: Philippines, 12 mm. Appears to be undescribed.

Elongate False Ear Shell *Stomatella impertusa* IP: Red Sea, Seychelles to Japan and Australia, 25 mm.

Dinghy False Ear Shell *Stomatella lintricula* WP: Japan to Philippines, Indonesia and Guam, 12 mm. Thin, oblong shell with convex back, easily recognized by its very elongate shape. Left: © D. Burdick. ID tentative.

Flattened Stomatella *Stomatella planulata* IWP: E. Africa to Japan, Philippines, Australia, 30 mm. Flattened oval shell. Dark olive with white dots or finely variegated and marbled with grey and olive-brown. ID tentative.

Caledonian Stomatella *Stomatella caledonica* SP: New Caledonia (endemic), 15 mm.

Golden Stomatella *Stomatella* sp. WP: New Guinea, 11 mm. Possibly undescribed. ID tentative.

Lattice Stomatella *Stomatella* sp. WP: New Guinea, 10 mm. Possibly undescribed. ID tentative.

Swollen Stomatella *Stomatia phymotis* IP: Red Sea, E. Africa to Japan, Australia and Marshalls, 35 mm.

Stomatella snails are fast-moving small nocturnal animals that prefer to feed on algae. They are cryptic animals with depressed, not perforated shell. Of nearly two dozen described species, most were described in the 19th century, and this implies problems with visual identification.
To inexperienced divers and underwater photographers, they resemble slugs, not snails.
a-e: *Stomatella* sp, f-i: *Stomatolina* sp.
Photos: a-d, f-i -©JJ. Identification tentative.

TROCHIDAE

Common Button Top *Umbonium vestiarium* IP: E. Africa, Philippines, to Australia, 15 mm. Small glossy shell, lenticular in shape, variable in pattern and coloration, in shades of grey, brown, olive green, pink. © Ria Tan.

LIOTIIDAE

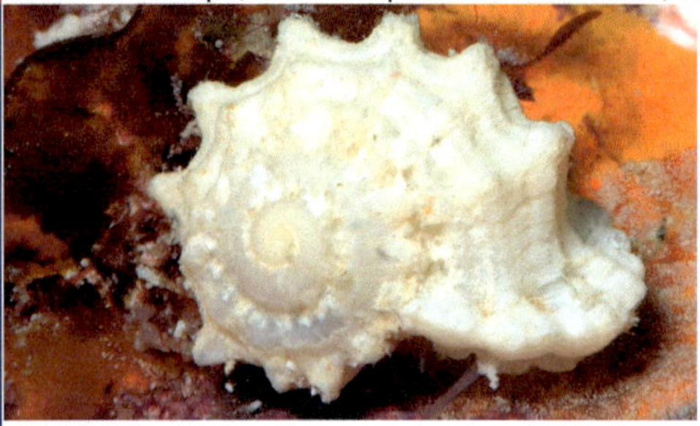

Divided Dentarene *Dentarene loculosa* WCP: Japan to Taiwan, Philippines and Marshalls, 14 mm. Small shell, white in color with crenulate whorls. Translucent white animal, orange-margined operculum. ©JJ.

Peron's Liotia *Liotinaria peronii* WCP: Australia, Fiji, Marshalls, 20 mm. Shell is white in color, whorls with prominent axial ribs and perforate depressions between them. Animal white with red striations. ©JJ.

ANGARIIDAE

Knotty Angaria *Angaria nodosa* IWP: India to Japan, Philippines and Marshalls, 55 mm. Cryptic animals, hard to notice due to heavy encrustation and irregular shape. Animal reddish, with white head. ©JJ.

HALIOTIDAE

Glistening Abalone *Haliotis glabra* WP: Philippines to Indonesia and Australia, 55 mm. The oval, depressed shell is marked with arrow-shaped olive spots on a lighter ground and 6-8 small perforations.

Donkey's Ear Abalone *Haliotis asinina* IWP: Andaman Sea to Japan, Australia and Palau, 121 mm. Outside of shell olive green or brown in color, with various, often triangular patches of green or cream and 5-7 open holes.

Oval Abalone *Haliotis ovina* IWP: Phuket to Marshalls, 70 mm. Ovate shell with holes on tubular projections, giving the spire a coronate appearance, last 4-6 holes open. Greenish with radiating cream stripes. ©SJ.

Lovely Abalone *Haliotis clathrata* IP: E. Africa to Japan, Samoa and Marshalls, 60 mm. Ovate shell, red to reddishbrown in color, holes on tubular projections, the last 4-5 open. ©JJ.

Canaliculate Abalone *Haliotis parva* WP, 60 mm. Red shell with a strong rib, parallel with the row of holes.

Red Abalone *Haliotis* sp. WP: Philippines, 20 mm. Close to *H. clathrata*, but have smoother shell. ID tentative.

Mouse Keyhole Limpet *Diodora mus* IP: India to Japan, Australia, Palau and Marshalls, 20 mm. Shell conical, with a hole at the apex. White with green, brown or red radiating stripes. Animal white with brown spots ©JJ.

Reeve's Keyhole Limpet *Diodora octagona* IWP: Thailand to Philippines and Marshalls, 20 mm. ©JJ.

Tall Keyhole Limpet *Diodora* sp. CP: Marshalls, 14 mm. Similar to *D. mus*, but taller. ©JJ. ID tentative.

Exceptional Notch-Limpet *Emarginella eximia* WCP: Japan to Australia and Marshalls, 10 mm. White perforate-looking shell with a notch on the anterior margin. Animal white. ©JJ.

Incised Notch-Limpet *Emarginella incisura* IP: India to Japan, Philippines, Marshalls, possibly Maldives (unconfirmed), 35 mm. White shell with radial ribs and curved notch on the anterior margin. Animal white. ©JJ.

Pretty Notch-Limpet *Emarginula concinna* WCP: Japan, Philippines, Marshalls, 10 mm. White conical shell with a straight notch on the anterior margin. Animal white. ©JJ.

Pretty Notch-Limpet *Emarginula concinna* (cont'd) dorsal view with depressions between radial ribs. ©JJ.

Wedged Notch-Limpet *Emarginula* sp. CP: Marshalls, 10 mm. Shell with a wedged notch. ©JJ.

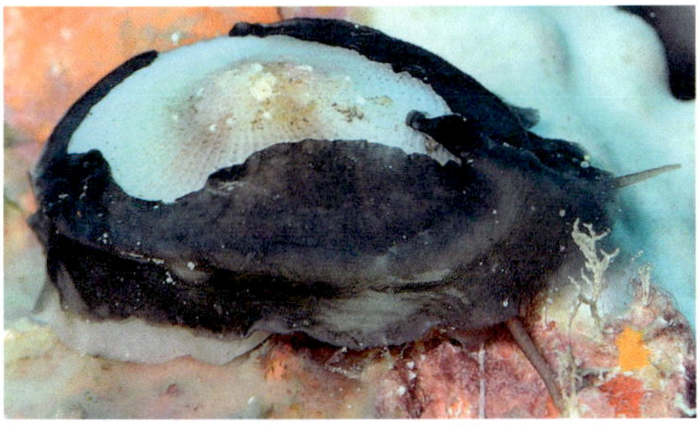

Flat Notch-Limpet *Emarginula* sp. CP: Marshalls, 24 mm. Shell with a straight notch, black mantle. ©JJ.

Flat Notch-Limpet *Emarginula* sp. CP: Marshalls, 18 mm. Shell with a straight notch, translucent mantle.

Spotted Notch-Limpet *Emarginula* sp. CP: Marshalls, 17 mm. Shell with a straight notch, translucent spotted mantle. Close to *E. concinna*, but shell profile and notch are different. ©JJ. ID tentative.

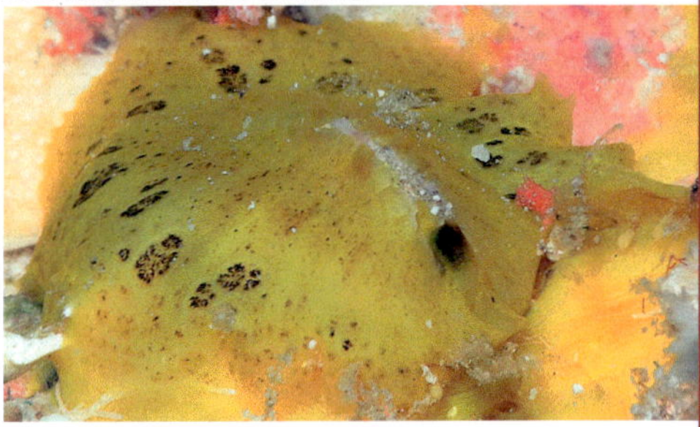

Siboga Notch-Limpet *Emarginula sibogae* CP: Marshalls, 20 mm. Shell cream in color, with a slightly wedged notch, black-spotted mantle. ©JJ. ID tentative.

Notch-Limpet *Emarginula* sp. CP: Marshalls, 15 mm. Shell with a notch, mantle with elevated folds. ©JJ.

Elephant Snail *Scutus antipodes* SP: Australia, New Zealand, Tasmania, 153 mm. © Uwe Weinreich.

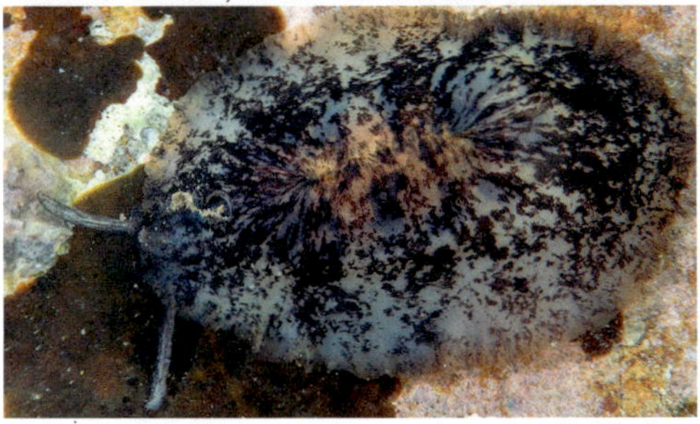

Chinese Shield Limpet *Scutus sinensis* WP: Thailand to Japan and Marshall Islands, 48 mm. Large slug-like limpet, conical shell with elevation in the middle (photo on the right). Nocturnal, feeds on algae. ID tentative.

KEYHOLE LIMPETS - FISSURELLIDAE

Hoof Shield Limpet *Scutus unguis* IWP: Red Sea, E. Africa, Japan to Australia, Marshalls, 60 mm. Large sluglike limpet, conical shell with elevation in the middle (photo on the right). Dark grey, translucent mantle.

Pointed Narrow Slot Limpet *Macroschisma cuspidatum* WP: Japan to Philippines, 20 mm. Small cap-like shell with long tubular siphon extending through it. ©JJ.

CHILODONTAIDAE

Blackish Margarite *Euchelus atratus* IWP: Red Sea, Reunion, Philippines to Singapore and Vanuatu, 21 mm. Shell with spiral sculpture and convex whorls, round aperture with dark maculation along the margin. © Ria Tan.

Blackish Margarite *Euchelus atratus* (cont'd) Shell with thick layer of periostracum. Subtidal rocks.

Spotted Granata *Granata maculata* WCP: Philippines to Australia, Tuamotu, 14 mm. © David Massemin.

Elegant Granata *Granata lyrata* WCP: Japan to Philippines and Marshalls, 15 mm. Small thin shell with spiral sculpture and convex whorls, white with pink-brown maculation on spiral cords. ©JJ.

Modest Sundial *Architectonica modesta* IWP: E. Africa to N. Caledonia, 55 mm. Depressed cone-shaped shell, inflated whorls. Frequently confused with *A. perspectiva*, (below) which has no spiral groove in the midrib-area.

Clear Sundial *Architectonica perspectiva* IP: Red Sea to Hawaii, 60 mm. Depressed cone-shaped shell, inflated whorls. White with variegated dark-brown and grey-brown bands. White tentacles with 2 black stripes each.

Cousin Sundial *Architectonica consobrina* Philippines, 36 mm, depth 250-370 m. © E.G de Suduiraut.

Areolate Sundial *Heliacus areola* Red Sea, E. Africa to Hawaii, 18 mm. White with pattern of square blotches.

Woven Sundial *Heliacus implexus* IP: Red Sea, E. Africa, Japan, Australia and Hawaii, 15 mm. Small depressed cone-shaped shell, white to yellowish with tan or brown blotches or flames. Feeds on *Palythoa* zoanthids. ©JJ.

Channeled Sundial *Heliacus infundibuliformis* IP: Reunion to Marshalls, 16 mm. Small depressed cone-shaped shell. Tan, reddish brown or dark brown with very thick, translucent yellowish periostracum. ©JJ.

Trochus-Like Sundial *Heliacus trochoides* IP: E. Africa to Maldives, Australia, Hawaii, Marshalls, 16 mm. Highspired cone-shaped shell. White with orange lines in grooves between spiral ribs. Animal white. ©JJ.

Variegated Sundial *Heliacus variegatus* IP: Red Sea, E. Africa to Philippines, Australia and Marshalls, 21 mm. Depressed or rounded cone-shaped shell. Bluish white with pattern of dark-brown flames. Left: ©JJ.

Sharp-Keeled Sundial *Psilaxis oxytropis* IP: Red Sea, E. Africa to Japan, N. Caledonia and Hawaii, 19 mm. Moderately depressed or even high-spired shell. Generally yellowish-brown, sometimes with lighter areas. ©JJ.

Sharp-Keeled Sundial *Psilaxis oxytropis* (cont'd)
Feeds on hard coral polyps and anemones. ©JJ.

Pale Sundial *Psilaxis* sp.©JJ CP: Marshalls, 14 mm.
Undescribed, but may be a young *P. radiatus*. ©JJ.

Radiant Sundial *Psilaxis radiatus* IP: Red Sea, E. Africa to Japan, Palau and Hawaii, 26 mm. cone-shaped shell whorls inflated, smooth and glossy. White with spiral brown band and regular axial flames. ©JJ.

Needle Pyram *Longchaeus acus* IP: E. Africa, Japan, Philippines to Australia and Fiji, 67 mm. White or yellowish shell with spiral bands of brown maculations.

Island Pyram *Longchaeus insularum* IP: Maldives to Japan, Marshalls and Hawaii, 30 mm. ©JJ.

Tower Pyram *Longchaeus turritus* IP: E. Africa to Marshalls, 24 mm. White with network of orange lines.

Sulcate Pyram *Longchaeus maculosus* IP: Red Sea, E. Africa to Philippines and Marshalls, 50 mm. Elongate fusiform shell, white to yellowish, with spiral bands of brown spots.

Auriscati Pyram *Otopleura auriscati* IP: Red Sea, Madagascar to Japan, Indonesia and Marshalls, 30 mm. Elongate fusiform shell, white in color, with spiral bands of brown maculations. ©SJ (left photo).

Ringed Pyram *Pyramidella dolabrata* IP, 35 mm. White shell with brown and dark blue spiral bands.

Needle Pyram *Styloptygma aciculina* WCP: N. Caledonia to Tahiti, Marshalls, 7mm. © C. Sylvester.

Peron's Onch Slug *Peronia peronii* IP: Mauritius to Philippines, Timor and Marshalls, 250 mm. Air-breathing sea slug, a shell-less marine pulmonate gastropod. Nocturnal, oval, bumpy animal, found on intertidal rocks. ©SJ.

Grey-Footed Mangrove Onch Slug *Platevindex* sp. WP: Singapore, 30 mm. Eyes on long thin stalks. Generally grey or brown. The underside grey, narrow foot with horizontal ridges. © Ria Tan.

Brown Onch Slug *Platevindex* sp. WP: Philippines, 50 mm. Intertidal rocks, mangrove area.

Pimply Onch Slug *Onchidiidae* sp. WP: Singapore, 50 mm. Eyes on on short stalks. © Ria Tan.

Smooth Onch Slug *Onchidiidae* sp. WP: Singapore, 30 mm. Dark grey or brown, smooth skin with small regular bumps. Eyes on long thin stalks. © Ria Tan.

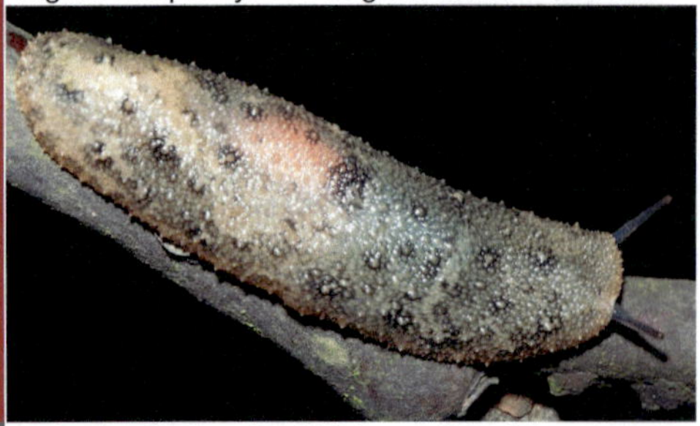

Long Onch Slug *Onchidiidae* sp. WP: Singapore, 30 mm. Mangroves, eyes on long stalks. © Ria Tan.

Ornate Onch Slug *Onchidium* sp. WP: Singapore, 50 mm. Skin with bumps and spikes. © Ria Tan.

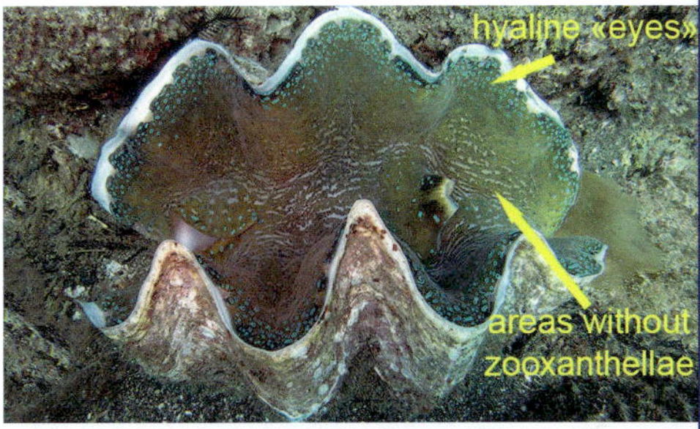

Giant Clam *Tridacna gigas* IP, 137 cm. Smooth shells with no ribbing and four or five vertical folds. Unlike other Tridacna species, incurrent siphon bears no tentacles. Areas with no zooxanthellae cells in the central axis are often marked with a white longitudinal lines. Mantle olive-brown or greenish. Iridescent blue ocelli, hyaline light-sensitive organs ("eyes") are concentrated near the edge of the mantle. In many regions giant clams are becoming endangered due to intensive exploitation. Photo: "f","h","i" ©JJ, photo "g": ©SJ.

Fluted Giant Clam *Tridacna squamosa* IP: Red Sea, South and East Africa to Japan, Australia, and Marshall Islands, 40 cm. **Identified by its large leaf-like (or spoon-like) scutes.** Live free-living, not embedded into a corals, incurrent siphon with distinct tentacles. Mantle greenish to brown or purple in linear or spot-like patterns. Gets a major portion of its nutrition from symbiotic zooxanthella, spreading out during the day its mantle tissue so that the algae receive the sunlight they need to photosynthesize. Photo: "e"-"j" ©JJ.

Small Giant Clam *Tridacna maxima* IP: Red Sea, East Africa to Maldives, Thailand, Japan, Australia, Polynesia, 35 cm (41.7 cm - largest individual collected). Juveniles are fully embedded in the reef, adults embedded only partially. Identified by its close-set scutes on the upper valves, the neat rows of tightly spaced hyaline organs ("eyes") along its mantle margin, and its brightly colored and mottled mantle (usually blue, green and brown). Rarely dwell beyond a depth of 10 m. Photo: "d","h" "j" ©JJ, "b","c","e","f","g","h","i" ©SJ.

Boring Clam *Tridacna crocea* IP: Seychelles to India, Japan, Palau, Solomons, Fiji and Australia, absent in Marshalls, 150 mm (the smallest of tridacnine species). A rock borer, that embeds its entire body into the reef.

Boring Clam *Tridacna crocea* (cont'd) The mantles are brightly colored, exhibiting various shades of blue, green, purple, white and brown. Scutes are typically eroded away by the natural burrowing activities.

Longest Giant Clam *Tridacna elongatissima* IO: Mozambique, Kenia, Reunion, 335 mm. Distinct from similar *T. maxima* by its green margin mantle and larger number of radial folds (6–7 vs. 4–6). © D. Goebbels.

Longest Giant Clam *Tridacna elongatissima* (cont'd) described in 1856, later it has been synonymized with either *T. maxima* or *T. squamosa* until resurrection in 2020 (Fauvelot et al 2020). © Dawn Goebbels.

Scaly Red Sea Clam *Tridacna squamosina* IO: Red Sea, 40 cm. Shell with 5-7 radial folds. Found exclusively in very shallow water. Mantle with network of fine lines, green rim and wart-like protrusions. © Irina Sevriuk.

Devil Clam *Tridacna mbalavuana* (former *T. tevoroa*) SP: Australia, Fiji, N. Caledonia, 57 cm. Warty mantle does not extend beyond shell margins. Inhabits deep waters (>20 m). Photo (left, right) © J. Adams, © J. Tiavouane.

Noah's Giant Clam *Tridacna noae* WP: Japan, Philippines to Australia and Micronesia, 50 cm. Distinguished from *T. maxima* through the presence of teardrop-shaped markings on the mantle, typically bounded by white.

Smooth Giant Clam *Tridacna derasa* WP: Cocos to Tonga, China to Australia, 60 cm. © Yves Trenevet.

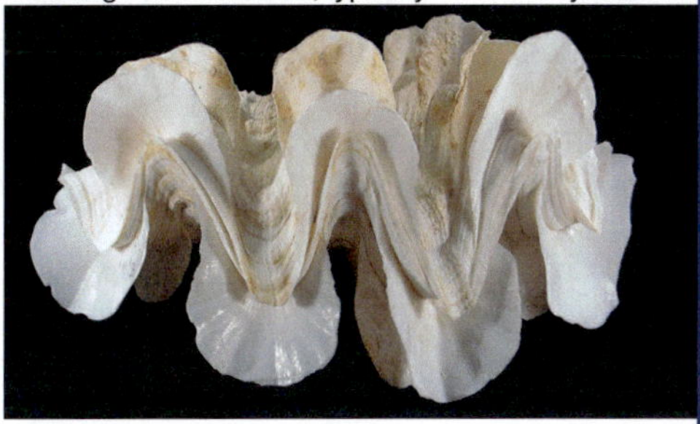

Rosewater's Giant Clam *Tridacna rosewateri* IO: Saya de Malha Bank, 200 mm. Thin walls. © B. Sirenko.

Horse's Hoof Clam *Hippopus hippopus* WP: Indonesia to Japan, Kiribati, Tonga, Australia, 50 cm. Thick shell with strong radial ribbing. Mantle lack hyaline organs ("eyes"), does not extend beyond shell margins. ©SJ.

China Clam *Hippopus porcellanus* WP: Indonesia, Malaysia, Philippines, New Guinea, 41 cm. Rounded shell, mantle grey or brown. Unlike *H. hipposus*, incurrent siphon with prominent papillae. © J. Adams.

Double-Rayed Cockle *Acrosterigma biradiatum* IP: Red Sea, E. Africa to Philippines, Australia and Marshalls, 50 mm. They have a short siphon and can use their long, strong, foot to jump from predators. ©JJ.

True Heart Cockle *Corculum cardissa* IP: Red Sea, Mauritius to Japan, Marshalls, Polynesia and N. Caledonia, 80 mm, usually to 50 mm. Strongly compressed shell, found in sandy bottoms or on *Halimeda* plants. ©JJ.

True Heart Cockle *Corculum cardissa* (cont'd) Filter feeder, feeds on plankton extracted from the water. Live in symbiosis with dinoflagellates in the genus *Symbiodinium*. Collected in some areas. ©JJ.

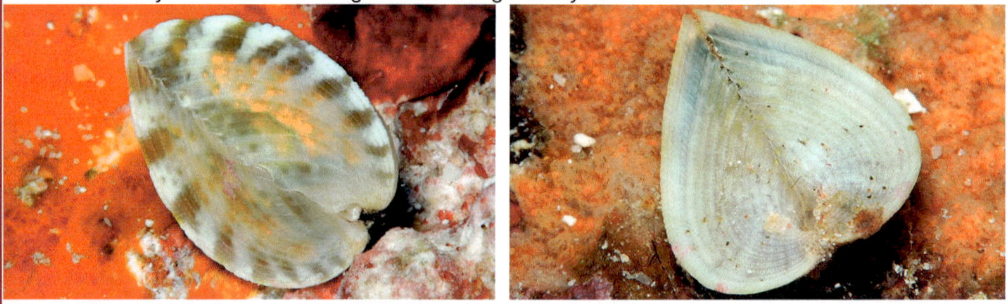

Lorenz' Heart Cocle *Corculum lorenzi* WP, Philippines to Marshall Islands, 51 mm. Smooth sided, spineless shell, close to *C. cardissa*. ©JJ.

Pacific Strawberry Cockle *Fragum unedo* IP: Mauritius, Sri Lanka to Japan, Australia and Melanesia, 65 mm. White or cream shell, about 27 radial ribs at each valve, with widely spaced red scales. © Ria Tan.

White Strawberry Cockle *Fragum fragum* IP: Red Sea, East Africa to Japan, Marshalls, Polynesia, Australia, 40 mm. White shell, about 32 radial ribs at each valve, with close spaced whitish or pale yellow scales. ©JJ

White Strawberry Cockle *Fragum fragum* (cont'd) Found on sand bottom to a depth of 20 m. Lives buried in sand, extending its siphons to the surface to draw in water in order to filter feed and breathe.

Lyrate Cockle *Lyrocardium lyratum* Red Sea, E. Africa to Japan, Marshalls and Australia, 50 mm. Smooth rounded shell, red or white with red clouds. Note the strong foot, preparing to jump (photo on the right) ©JJ.

Winner Cocle *Freneixicardia victor* IWP: Mauritius to Maldives, Marshalls, 60 mm. Deep water (>60 m.) ©SJ.

Orbit Cockle *Vasticardium orbita* WCP: Philippines to Hawaii, 93 mm. Shell w/purple internal margins. ©JJ.

Elongated Cocle *Vasticardium elongatum* IP: E. Africa, Philippines to Australia, Fiji and Marshalls, 136 mm. Large shell, elongate and inflated. Exterior cream with various shades of brown-yellow to purple.

Scaly Galeommatid *Galeommella utinomii* IP: Reunion to Japan, 7 mm. Shell transparently white, elliptical oval in shape, with distinct radial ribs. Mantle with red-orange spots, covers the shell. © Johji Nishio.

Semper's Galeommatid *Devonia semperi* WP: Japan, Philippines, 20 mm. Symbiotic with holothurians.

Elegant Galeommatid *Scintilla nitidella* WP: Japan to N. Caledonia, 13 mm. ID tentative.

Cuvier's Galeommatid *Scintilla cuvieri* WP: Philippines, Hong Kong, Indonesia,19 mm. Inflated shell, bronze to bright yellow in color. Associated with a number of invertebrates, including ophiuroids and alpheid shrimps.

Philippine Galeommatid *Scintilla philippinensis* WP: Philippines to Singapore, 20 mm. Inflated shell with weak hinge. Associated with peanut worms. Mimics 2 kinds of animals: nudibranchs and crabs. © Arthur Anker.

Red-Tipped Galeommatid *Scintilla* sp. WP: Singapore, 8 mm. © Arthur Anker. ID tentative.

Yellowish Galeommatid *Scintilla* sp. WP: Singapore, 9 mm. © Arthur Anker. ID tentative.

Fragile Galeommatid *Scintilla* sp. SP: New Caledonia, 10 mm. Shell transparently yellowish, elliptical oval in shape, smooth. Shallow intertidal, under rocks, host unknown. NHMN Koumac-2018 expedition.

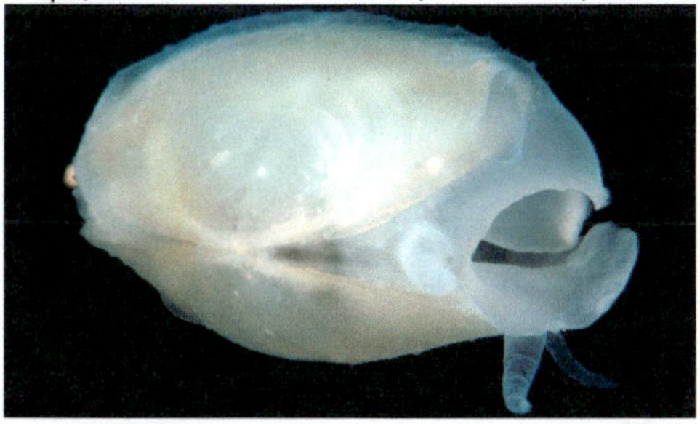

Smooth Galeommatid *Scintilla* sp. WP: Philippines, Indonesia, 12 mm. Shell transparently white, elliptical oval in shape, smooth. Shallow subtidal, under rocks, host unknown. ID tentative.

Tuberculated Galeommatid *Scintilla* sp. WP: N. Guinea to Philippines, 10 mm. Mantle w/orange-tipped tubercles.

Fuzzy Galeommatid *Scintilla* sp. WP: Indonesia, 12 mm. Shallow rocks, associated with brittle stars.

Mammillate Galeommatid Scintilla sp. SP: New Caledonia, 10 mm. Shell transparently white, elliptical oval in shape. Mantle totally covered by rounded tubercles. NHMN Koumac-2018 expedition.

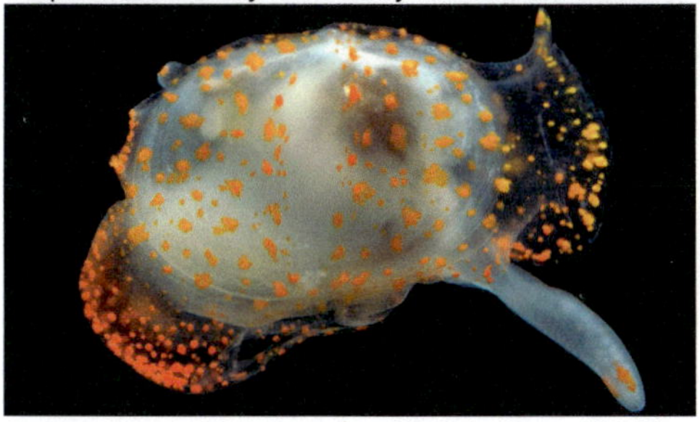

Orange-Spotted Galeommatid *Scintilla* sp. SP: New Caledonia, 9 mm. NHMN Koumac-2018 expedition.

Red-Shell Galeommatid *Scintilla* sp. WP: Papua New Guinea, 8 mm. © Arthur Anker. ID Tentative.

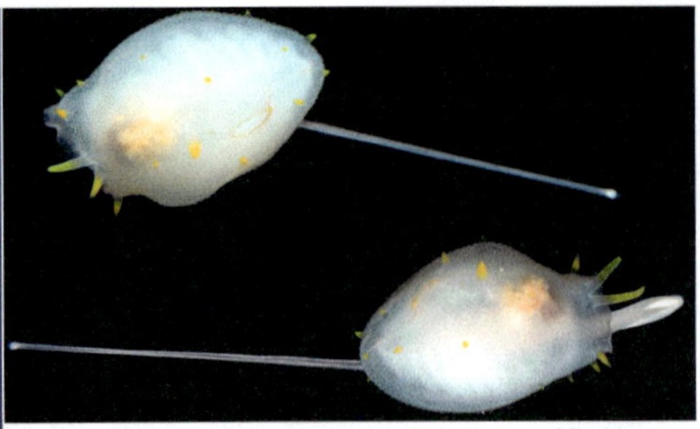

Yellow-Tipped Galeommatid *Scintilla* sp. SP: New Caledonia, 10 mm. NHMN Koumac-2018 expedition.

Brown-Spotted Galeommatid *Scintilla* sp. WP: New Guinea to Philippines, 11 mm. © Arthur Anker. IT.

Jukes' Galeommatid *Scintilla* cf *jukesi.* WP: Philippines, Australia, 12 mm. © E.G de Suduiraut.

Red-Spotted Galeommatid *Scintilla* sp. CP: Micronesia, 14 mm. White mantle, red-tipped tubercles ©JJ.

White-Tipped Galeommatid *Scintilla* sp. WP: Papua N. Guinea,11 mm. © Arthur Anker. ID Tentative.

Bloody-Shell Galeommatid *Scintilla* sp. WP: Papua New Guinea, 9 mm. © Arthur Anker. ID Tentative.

Open Galeommatid *Ephippodonta* sp. IO: Red Sea, 9 mm. Valves always open at 180°. © Arthur Anker.

Ochetostoma's Bivalve *Borniopsis ochetostomae* IP, 10 mm. Spoon worm commensal. NHMN 2018.

Anomal Galeommatid *Varotoga anomala* WP: Thailand to Australia, 9 mm. The mantle is covered with tentacles of varying length. The posterior margin of the left and right inner mantle lobes is yellowish. NHMN Koumac-2018.

Planetary Venus *Callista planatella* WP: Philippines to Australia and New Caledonia, 83 mm.

Script Venus *Circe scripta* IWP: Red Sea, E. Africa, India, Japan to Australia, 83 mm.

Ribbed Venus *Gafrarium pectinatum* IWP: Red Sea, E. Africa, Philippines to Australia, 43 mm.

Hieroglyph Venus *Lioconcha hieroglyphica* WCP: New Guinea to Marshalls and Hawaii, 43 mm. ©SJ.

Zigzag Venus *Lioconcha castrensis* IWP: Red Sea, Reunion, Japan to Indonesia, 58 mm. Outside with large dark, V-shaped markings or zigzag lines. Right: *L. castrensis* f. *lorenziana*, formerly treated as a separate species.

Ornate Venus *Lioconcha ornata* IP: Red Sea, Australia to Philippines and Japan, 47 mm.

Philippine Venus *Lioconcha philippinarum* WCP: Japan to Indonesia, Marshalls, 66 mm. ©SJ.

Tiger Venus *Lioconcha tigrina* IWP: Red Sea, E. Africa to Indonesia, Philippines and Polynesia, 39 mm. They can use their long, strong, foot to jump from predators (left photo: preparing to jump).

Embossed Venus *Globivenus toreuma* IP: Red Sea, E. Africa, Japan to Australia and Marshalls, 50 mm. Roundly ovate shell with coarse growth lines, white with reddish clouds. ©JJ.

Lattice Venus *Periglypta exclathrata* IP: E. Africa to Japan, Philippines, Marshalls, 65 mm. ©JJ.

Reticulated Venus *Periglypta reticulata* IP: Red Sea, E. Africa to Japan, Australia, Marshalls, 72 mm. ©SJ.

Lamellated Venus *Placamen lamellatum* WP: Thailand to Japan, New Guinea and Australia, 40 mm. Shell is white in color with brown bands. Valve with lamellae, shaped like a bird's wing.

Shaved Jewel Box *Chama asperella* IP: Red Sea, E. Africa,Philippines to Australia, 47 mm. ©SJ.

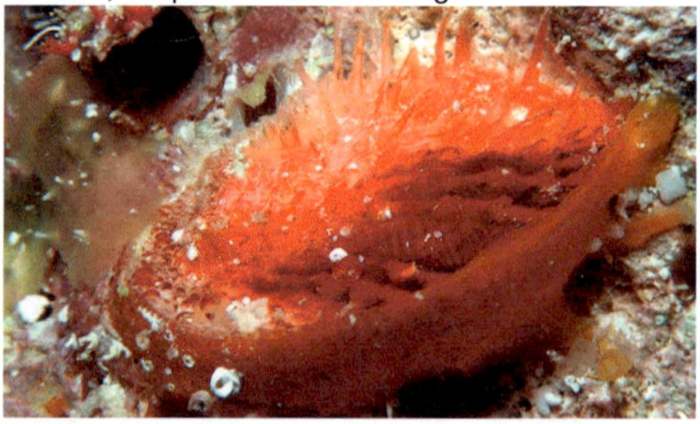

Reflexed Jewel Box *Chama pacifica* IP: 80 mm. Upper valve with radial rows of spines or scales. ©SJ.

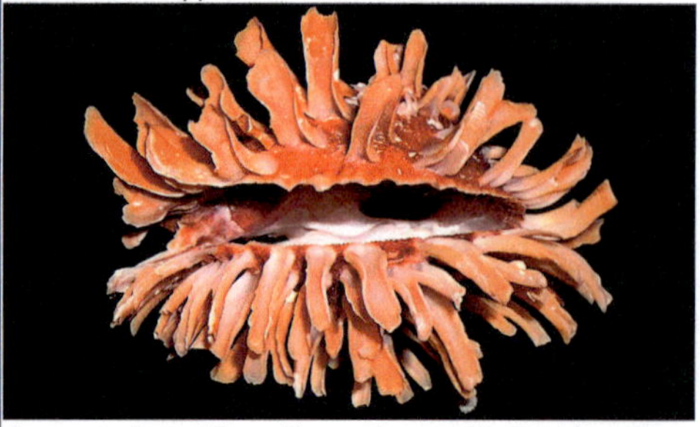

Lazarus Jewel Box *Chama lazarus* IP: Red Sea, East Africa to Japan, Australia and Melanesia, 140 mm. Outer surface with long, leaf-like expansions. Often collected for its beautiful shell. © Richard Parker.

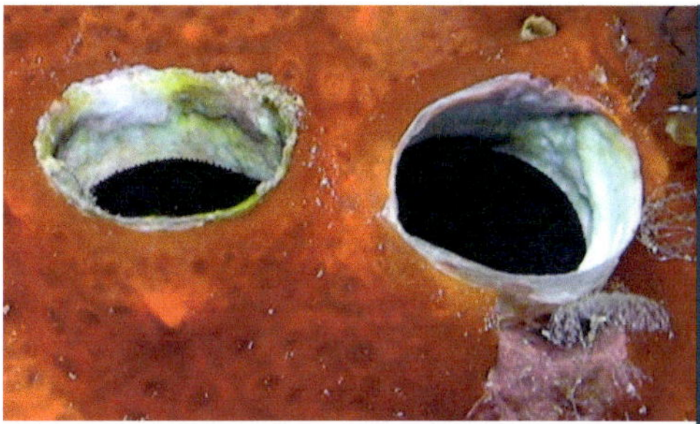

Watering Pot Shell *Clavagella* sp. IP: similar species from Red Sea to Hawaii. Family contains weird bivalves, encased in a tube, with an anterior 'watering pot', perforated with tiny holes. Photo (left, right) ©JJ, ©SJ.

Antique Ark *Anadara antiquata* IP: Red Sea, E. Africa to Polynesia and Marshalls, 105 mm. Solid ovate shell with about 40 radial ribs at each valve, radial ribs with a narrow median grooves. Photo (left, right) ©JJ, ©SJ.

Ventricose Ark *Arca ventricosa* IP: Red Sea, E. Africa to Polynesia and Marshalls, 90 mm. Elongated shell, fixed on rocks, corals, usually encrusted with marine growths. Anterior half creamy white with zebra-like brown stripes. ©SJ.

Painted Ark *Arca* sp. CP: Marshall Islands, 25 mm. Small shells, found under rocks on lagoon reefs. White with orange clouds. Photo on the left: shell encrusted with orange sponge. ©JJ.

Dark Mantle Ark *Arca* sp. WP, 150 mm. Several similar species, ID needs close examination.

Blood Cockle *Tegillarca granosa* IP: E. Africa to Polynesia, 90 mm. About 18 stout ribs. © Ria Tan.

Electric File Shell *Ctenoides ales* WCP: Japan to Australia and Marshalls, 93 mm. Red mantle with iridescent blue margins. Flashing-light display comes from reflection of the ambient light from the white reflective tissue.

Annulate File Shell *Ctenoides annulatus* IP: Red Sea, Madagascar, Japan to Australia and Marshalls, 41 mm. Moderately compressed shell with network pattern, tan to brownish in color. Found under shallow rocks. ©JJ.

Annulate File Shell *Ctenoides annulatus* (cont'd) Numerous tubercles tan with darker annulations. ©JJ.

Common File Shell *Lima* sp. WP: Singapore, 20 mm. Ribs without scales. © Tsu Suu Tan.

Common File Shell *Lima vulgaris* IP: Red Sea, E. Africa, Japan to Australia, Marshalls and Polynesia, 90 mm. Thick shell with coarse scaly ribs. Unattached specimens actively swim. Photos: (left, right) ©JJ, © U. Weinreich.

Orange File Shell *Limaria basilanica* WP: Japan, Philippines to Australia and New Caledonia, 42 mm. Compressed white shell, animal with pale orange tentacles of varying length. ID tentative.

Fragile File Shell *Limaria fragilis* IP: E. Africa, Maldives, Australia, Japan, Tahiti, 29 mm. Valves almost flat, intervals between main ribs with tiny scratches. Long, deep red tentacles mimic an anemone.

Fragile File Shell *Limaria fragilis* (cont'd) When autotomized, tentacle wriggle like worms.

Cuming's File Shell *Limaria* cf. *cumingii* WP, 22 mm. Small shell, orange mantle, translucent tentacles. IT.

Pale File Shell *Limaria hakodatensis* WP: Japan to Indonesia, 24 mm. White mantle with red bands.

Stormy File Clam *Lima nimbifer* IP: S. Africa, Australia and N. Caledonia, 90 mm. NHMN Koumac-2018.

Red-Margin File Shell *Limaria* sp. CP: Marshall Islands, 20 mm. Small shell, animal white with a tint of red developing near the ends of the tentacles and a red fringe to the internal cavity. ©JJ.

Bleached File Shell *Limaria* sp. CP: Marshall Islands, 20 mm. Similar to previous species, but lacks the red on the tentacles and at the edges of the body cavity. ©JJ.

LIMIDAE

Red-Margin File Shell *Limaria* sp. CP: Marshall Islands, 20 mm. Small elongated shell, reddish animal with fine banding on the tentacles. ©JJ. ID tentative.

LUCINIDAE

Fimbriate Clam *Fimbria fimbriata* IP: Thailand to Fiji and Marshalls, 104 mm. The shell is heavy and thick, with pronounced radial ridges (flattened and forming overlapping plates in adults). Photo (left, right): ©JJ, ©SJ.

MYTILIDAE

Tapering Clam *Lithophaga teres* IP: Red Sea, E. Africa, Maldives, Japan to Marshalls, 94 mm.

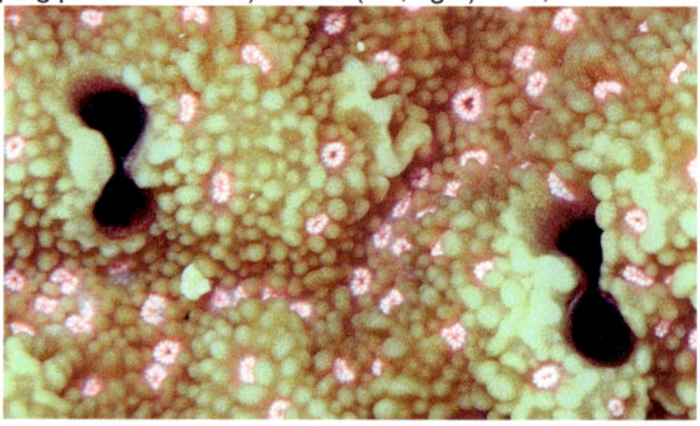

Dumbell Clam *Lithophaga* sp. WP. Forms dumbell shaped holes in coral. ©JJ. ID tentative.

HONEYCOMB OYSTERS - GRYPHAEIDAE

Honeycomb Oyster *Hyotissa hyotis* IP: Red Sea, E. Africa to Japan, Australia and Polynesia, 300 mm. Large shell, sharp and acute plicate margin reflecting the shape of the well-defined, regularly-spaced radial ribs.

Chinensis Honeycomb Oyster *Hyotissa sinensis* WCP, 150 mm. Large rounded shell, surface irregularly folded with radial ribs, margin with irregular wavy plications corresponding to rib positions. Yellow or deep purple. ©SJ.

Fingerprint Oyster *Alectryonella plicatula* IWP: E. Africa, Persian Gulf, Japan to Philippines and Indonesia, 120 mm. Outside of shell light to dark greyish brown in color, with 14-28 radial folds ending in a zigzag margin.

Leaf Oyster *Dendostrea folium* IP: Red Sea to Melanesia: 100 mm. Small to medium size, elongate oval, margin irregularly folded, radial ribs with rounded tops. Some authors consider *L. cristagalli* as synonym of *D. folium*.

Cock's Comb Oyster *Lopha cristagalli* IWP: Red Sea, E. Africa to Melanesia, 205 mm. Morphological features of oyster's shell may be exceedingly variable among individuals, making species identification sometimes very difficult.

Cock's Comb Oyster *Lopha cristagalli* (cont'd) About 4 - 8 very large, sharp radial folds on both valves. ©SJ.

Hooded Oyster *Saccostrea cuccullata* IP: Red Sea, E. Africa to Japan, Australia and Polynesia, 20 cm.

Blacklip Pearl Oyster *Pinctada margaritifera* IP: Red Sea to Persian Gulf, Australia, Japan and Polynesia. Outer surface dark grey with white radial stripes. Interior of the shell with black coloring along the margins. ©SJ.

Rayed Tree Oyster *Isognomon perna* IP: E. Africa to Japan, New Caledonia, Marshalls, Hawaii and Polynesia, 75 mm. Cream to brownish shell with recognizable pattern of darker brown rays. ©SJ.

Striped Wing Oyster *Pterelectroma physoides* IP: Red Sea, E. Africa, Japan to Indonesia, Marquesas, 27 mm. Associated with hydroids on sand/mud slopes.

Penguin Wing Oyster *Pteria penguin* IWP: Red Sea, E. Africa to Japan, Australia and Fiji, 300 mm. Large, dark brown to black shell, attached to rocks, corals, gorgonians. Collected for food and pearls.

Egyptian Wing Oyster *Pteria aegyptiaca* IO: Red Sea endemic, 100 mm. Shell with dark brown reticulation.

Reticulate Wing Oyster *Pteria* sp. WP: Indonesia, Philippines, 50 mm. Found on gorgonians.

Striped Wing Oyster *Pteria* sp. WP: Indonesia, New Guinea, 60 mm. Found on gorgonians.

Sponge Finger *Vulsella vulsella* IP: Red Sea to Marshalls, 104 mm. Lives embedded in sponges. ©SJ.

Comb Pen Shell *Atrina pectinata* IP: E. Africa to Japan, Australia and Marshalls, 370 mm. Large wedge-shaped shell, mantle with orange border and wavy yellowish lines, often interrupted. Photo on the left ©SJ.

Flag Pen Shell *Atrina vexillum* IP: Red Sea, Madagascar to Hawaii and Polynesia, 480 mm.

Black Mouth Pen Shell *Pinna* sp. WP: Indonesia, Philippines, 200 mm. Found into reef crevices.

Prickly Pen Shell *Pinna muricata* IP: Red Sea, E. Africa to Polynesia and Marshalls, 310 mm. Readily identified by rows of triangular spines. Like other family members, lives vertically imbedded in soft bottoms.

Large Pen Shell *Pinna* sp. WP: Philippines, Indonesia, New Guinea, 580 mm. Large dark-brown shell.

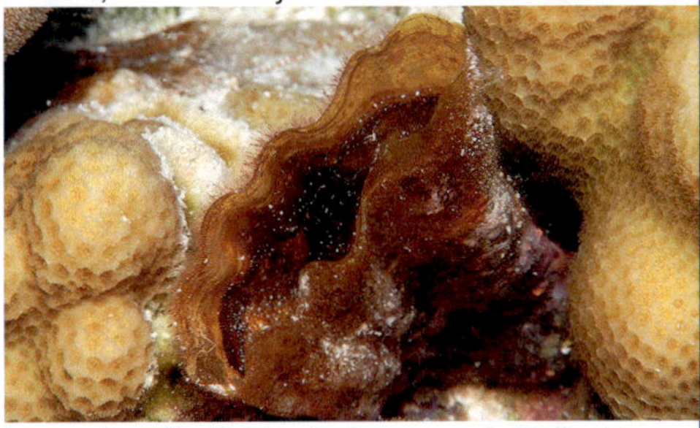

Baggy Pen Shell *Streptopinna saccata* IP: Red Sea, E. Africa to Philippines and Hawaii, 150 mm. ©JJ.

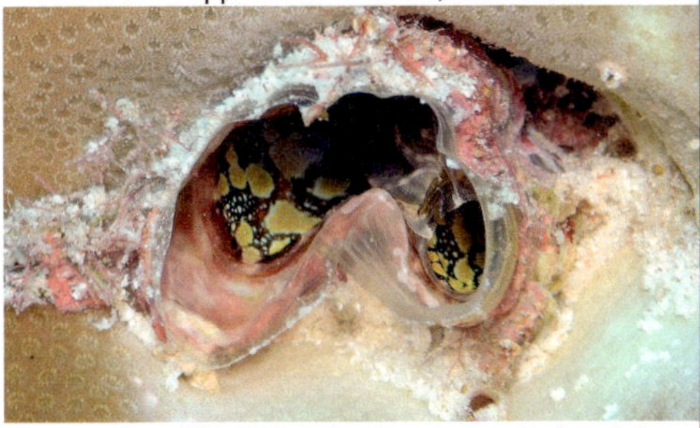

Baggy Pen Shell Streptopinna saccata (cont'd) Found growing out of crevices in coral or rocks. ©JJ.

Superb Scallop *Anguipecten superbus* WCP: Philippines, Japan to Marshalls, 60 mm. Medium sized shell, red with with white spots or bands, with 20-24 ribs. ©SJ.

Distant Scallop *Bractechlamys vexillum* WCP: Philippines, Indonesia to N. Caledonia, Marshalls, 60 mm. Hinge line is black on the inside, about 9 ribs. Free living. Able to swim about in jerky motion when disturbed. ©JJ.

Spectacular Scallop *Excellichlamys spectabilis* IP: Reunion, Maldives, Japan to Polynesia, 45 mm. Shell with brown and red dots and 1-3 more or less defined secondary ribs with scales between the primary ribs. ©JJ.

Effaced Scallop *Dentamussium obliteratum* WCP: N. Caledonia, Philippines to Marshalls, 66 mm. ©SJ.

Flatribbed Scallop *Decatopecten radula* WP: Japan to N. Caledonia, 94 mm. Ears with strong radial ridges.

Royal Cloak Scallop *Gloripallium pallium* IP: Red Sea, E. Africa to Thailand, Philippines, Australia, Marshalls and Polynesia, 78 mm. Shell with 12-14 ribs, similar *G. maculosum* has 7-9 rounded ribs. ©JJ.

Wedge-Shaped Scallop *Laevichlamys cuneata* WCP: Japan to N. Caledonia and Marshalls, 40 mm. Medium sized free living scallop with numerous fine ribs.

Nebulous Scallop *Laevichlamys* sp. CP, Marshalls, 36 mm. Close to *L. cuneata* but shell is smoother. ©JJ.

Punctate Scallop *Mimachlamys punctata* WP: East Africa to Indonesia, 50 mm. Shell with 22-26 ribs.

Wonderful Scallop *Mirapecten mirificus* IP: Reunion to Japan, Australia and Marshalls, 45 mm. ©SJ.

Molucca Scallop *Mirapecten moluccensis* WCP: Japan to Marshalls, 35 mm. Ribs with small nodules.

Rake Scallop *Mirapecten rastellum* IP: Red Sea to Philippines, Indonesia, New Caledonia, 52 mm. Easily identified by sharp scales on the ribs. Shell with red, yellow and white clouds. ©JJ.

Rake Scallop *Mirapecten rastellum* (cont'd) Shells are variable in color, usually encrusted with sponges. ©JJ.

Smooth Scallop *Mirapecten* sp. CP, Marshalls, 40 mm. Close to *M. rastellum*, but shell is smooth. ©JJ.

Flashing Scallop *Pascahinnites coruscans* IP: East Africa, Japan to Australia, Hawaii and Polynesia, 25 mm. The shell is variable in color, from white with brown spots to red, with 12-15 well formed primary ribs. ©JJ.

Pale Scallop *Pectinidae* sp. CP: Marshalls, 30 mm. Possibly a species of *Laevichlamys*. ©JJ. ID tentative.

Spotted Scallop *Pectinidae* sp. CP: Marshalls, 30 mm. Possibly a species of *Laevichlamys*. ©JJ. ID tentative.

High Scallop *Semipallium dringi* WCP: Philippines, New Caledonia, Marshalls, 50 mm. Round shell is variable in color with shades of red and yellow-brown, with a shagreen microstructure. ©JJ.

Iridescent Scallop *Pedum spondyloideum* IP: Red Sea, E. Africa to Japan, Australia, Marshalls, Polynesia, 90 mm. White shell with red-brown blotches. Occurs on a number of different kinds of living corals.

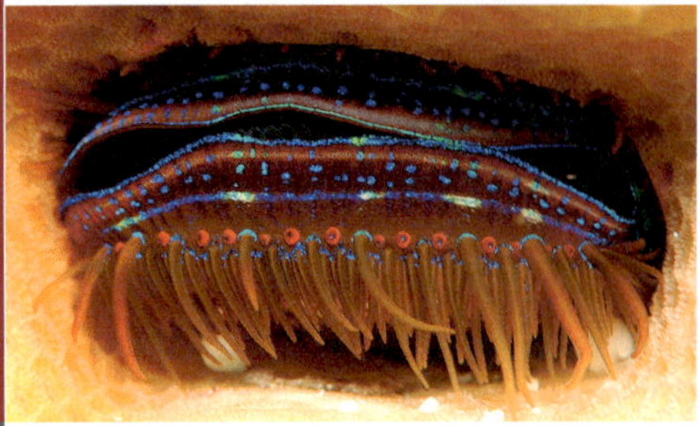

Iridescent Scallop *Pedum spondyloideum* (cont'd) Helps coral host to survive Crown of Thorns predation. Contact by starfish usually caused the scallops to generate powerful water jets, forcing starfish to move away.

Nude Thorny Oyster *Spondylus anacanthus* IP: E. Africa to Japan, Australia, Marshalls, 86 mm. Large rounded shell, usually coated with red sponge. Photo (left, right) ©SJ, ©JJ.

Butler's Thorny Oyster *Spondylus butleri* WCP: Japan, Philippines, Micronesia,100 mm. Shell with thick radial ribs, bearing strong flattened spines. Attached to rocks or dead corals. Photo (left, right): ©SJ, ©JJ.

Prickly Thorny Oyster *Spondylus echinatus* IP: Red Sea, East Africa to Philippines, Australia and Marshalls, 130 mm. Variable species in terms of spine development and coloration. ©JJ.

Prickly Thorny Oyster *Spondylus echinatus* (cont'd) The juvenile is white with small black patches. ©SJ.

Variagated Thorny Oyster *Spondylus variegatus* IWP: E. Africa to Philippines, Indonesia, 120 mm.

Cat Tongue Thorny Oyster *Spondylus linguafelis* CWP: Philippines to Solomons, Hawaii, Marshalls, 151 mm. Spiny shell, often encrusted. Photo on the left: colonized by *Nemanthus* sea anemones.©JJ.

Small Thorny Oyster *Spondylus* sp. CP: Marshalls, 30 mm. Cemented to the undersurfaces of rocks. ©SJ.

Bristled Thorny Oyster *Spondylus* sp. CP: Marshalls, 80 mm. Ribs with curved spines. ©SJ.

Scaly Thorny Oyster *Spondylus squamosus* IP: Mauritius to Phuket, Thailand, Japan, Philippines, N. Caledonia and Marshalls, 120 mm. Scaly radial ribs, ending with pale flattened spines. Photo (left, right): ©SJ, ©JJ.

Scaly Thorny Oyster *Spondylus squamosus* (cont'd) Left: juvenile. All spiny oysters are filter feeders. The adults live cemented to hard substrates. Right: adults, © E.G de Suduiraut.

Giant Variable Thorny Oyster *Spondylus varius* IP: Mauritius, Red Sea, Maldives to Japan, Australia, Marshalls and Polynesia, 225 mm. Found cemented reef slopes, they tend to clam up when approached.

Giant Variable Thorny Oyster *Spondylus varius* (cont'd) They have multiple eyes around the edges of their mantle, and relatively well-developed nervous system.

Britaev's Chiton *Acanthochitona britayevi* WP: Vietnam, 13 mm. © O. Savinkin.

Jewel Spiny Chiton *Acanthochitona* cf. *leopoldi* WP: Philippines, 15 mm.

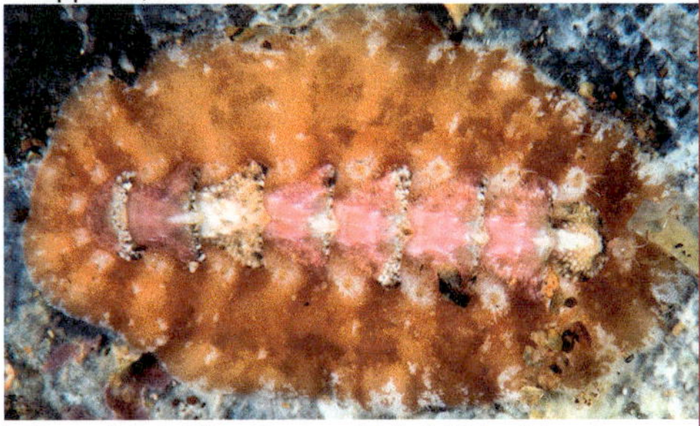

Velvety Chiton *Leptoplax* sp. WP: Philippines, 20 mm. Medium-sized chiton, trapezoidal pinkish plates are surrounded by orange-brown to tan girdle. Common on shallow rocks.

Jewelled Chiton *Acanthopleura gemmata* IWP: East Africa, Phuket, Philippines to Australia, Vanuatu, 70 mm. Large, oval shaped body with dimpled plate margins, a brown girdle and long spicules.

Vaillant's Chiton *Acanthopleura vaillantii* IO: Red Sea, Oman, mainland Yemen and Socotra, Persian Gulf, 50 mm. Large elongate oval body, valves are beaked and solid. Head valve is semicircular.

Spiny Chiton *Acanthopleura spinosa* WP: Philippines, Indonesia, Australia, 100 mm.

Balinese Chiton *Rhyssoplax baliensis* WP: Indonesia, 50 mm. Large, girdle with rounded triangular scales.

CHITONIDAE

Densilirata Chiton *Rhyssoplax densilirata* WCP: Philippines, Indonesia, Papua New Guinea, 52 mm. Variable in color from olive-green to greyish-brown, reticulated girdle, plates often with a dark wedge-shaped marks.

Densilirata Chiton *Rhyssoplax densilirata* (cont'd) Common on the shallow rocks in the Philippines.

Mauritanian Chiton *Rhyssoplax mauritiana* IO: East Africa to Seychelles, 40 mm. © Philippe Bourjon.

Lamellate Chiton *Lucilina lamellosa* WP: Vietnam and Philippines to N. Caledonia and Samoa, 40 mm. Coloration variable, of chestnut brown and greyish-green. Identified by the pattern of eye-like sensors on the plates.

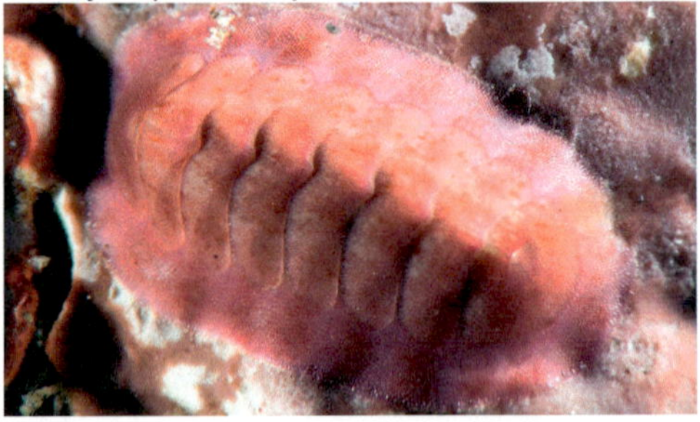

Hulule Chiton *Tegulaplax hululensis* IWP: Red Sea, E. Africa to Maldives, Vietnam, Fiji, also Mediterranean Sea, 32 mm. Plates and girdle are lilac in color.

CRYPTOPLACIDAE

Dawydoff's Chiton *Cryptoplax dawydoffi* WP: Vietnam to Philippines, Indonesia, 40 mm.

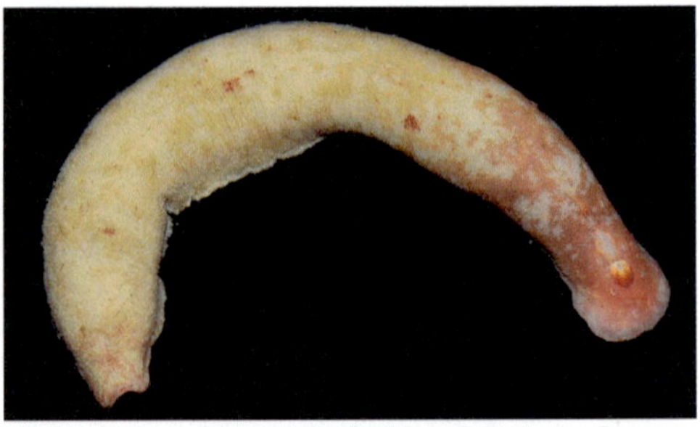

Eliot's Chiton *Cryptoplax elioti* New Caledonia, Samoa to Marshalls, 100 mm. Plates are much reduced.

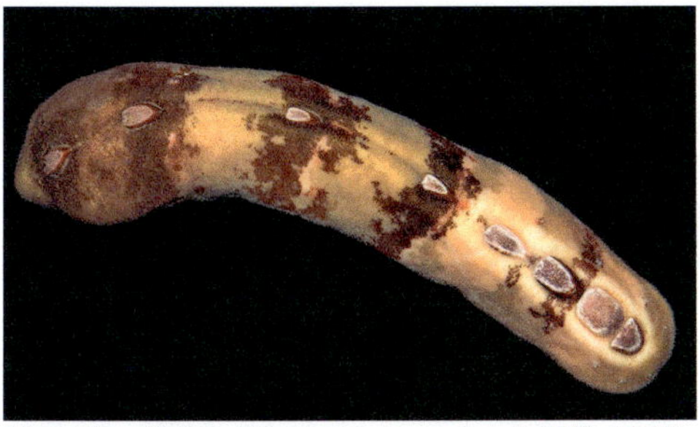

Slug-Like Chiton *Cryptoplax larvaeformis* IWP: Maldives and Thailand to Philippines, Indonesia, Fiji and Australia. 150 mm. Plates are reduced, flexible fast-moving animal. Nocturnal, hides under rocks during the day.

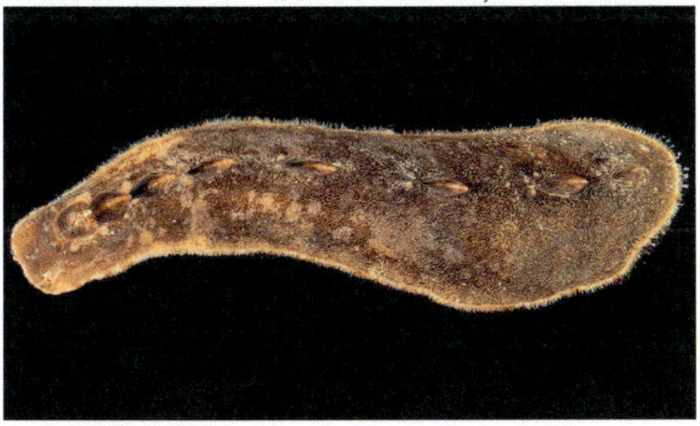

Caledonian Chiton *Cryptoplax larvaeformis* f. *caledonicus* New Caledonia, 120 mm.

Eye Chiton *Cryptoplax oculata* WP, 130 mm. Found on hard corals, retreats in the crevices when disturbed.

Adorned Chiton *Ischnochiton comptus* WP: Japan, Korea, China, Philippines: 25 mm. Variable in color, girdle scales are smooth and large. Found on intertidal rocks.

Bonin's Chiton *Ischnochiton* cf. *boninensis* WP: Philippines, 30 mm. Close to *I. boninensis* from Japan.

Alata Chiton *Stenoplax alata* IWP: India to Vietnam, Japan, China, 51 mm. Brownish girdle with dark spots.

Swollen Chiton *Schizochiton incisus* IWP: Seychelles, Chagos, Maldives, Philippines to Indonesia, Solomons and Australia, 50 mm. Easily recognizable by the deep V-shaped caudal sinus in the tail valve.

REEF ID BOOKSHELF

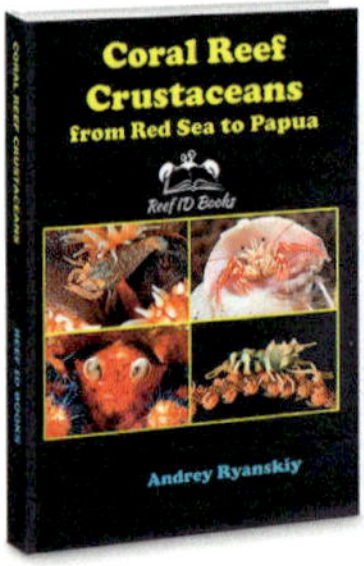

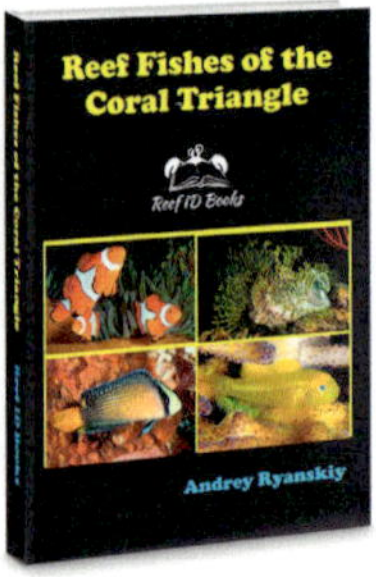

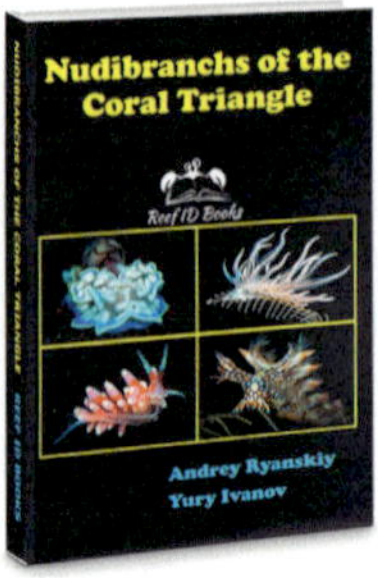

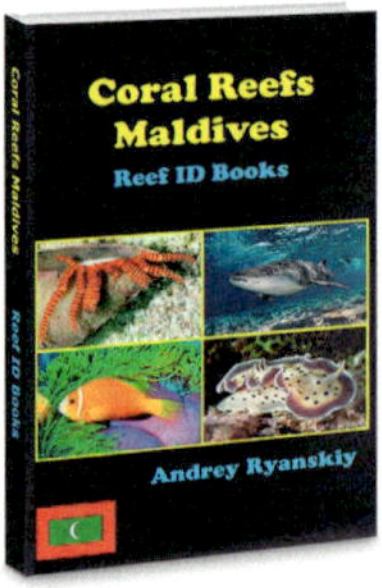

A project such as this would not be possible without significant input from numerous colleagues, photographers, marine biologists and citizen scientists for which author is most grateful (see Acknowledgment for details)

I am especially indebted to Jeanette Johnson and Scott Johnson, whose kindness, patience, love for marine life and deep knowledge inspired me to create this book and whose photographs made this project possible.

Thank you for taking time to read Living Seashells of the Tropical Indo-Pacific. If you enjoyed it, please consider telling your friends and posting a short review. Word of mouth is an author's best friend and much appreciated. Believe me, I read them all. Thank you!

Andrey Ryanskiy www.reefidbooks.com

Made in the USA
Las Vegas, NV
03 November 2026

57420364R00159